Digital Photogrammetry

Wilfried Linder

Digital Photogrammetry

A Practical Course

Fourth Edition

 Springer

Wilfried Linder
Geographisches Institut
Universität Düsseldorf
Düsseldorf
Germany

Additional material to this book can be downloaded from http://extras.springer.com.

ISBN 978-3-662-57063-0   ISBN 978-3-662-50463-5   (eBook)
DOI 10.1007/978-3-662-50463-5

Printed on acid-free paper

This Springer imprint is published by Springer Nature
The registered company is Springer-Verlag GmbH Berlin Heidelberg

Preface to the Fourth Edition

During the time since the last edition, the software development went on, and therefore it was necessary to adapt the text according to the actual program versions which were also tested to work properly under MS Windows 10. A few options which are obsolete nowadays were removed, some new options were added, handling of the software was simplified in several places, and so on. For the use of own images taken with a custom digital camera, a new chapter offers an easy way of lens calibration and explains how to get good results with a minimal work. Nevertheless, the general structure and the aims of this book remain unchanged: Step by step the reader is led through several tutorials to see and learn the basics of photogrammetric processing.

Like in the previous editions, not only the input data but also the intermediate and final results are presented, so that it is possible to skip parts of a tutorial and go on with a later state. The software and the data are no more delivered on CD-ROM but are available in a server: Just go to extras.springer.com and key in the ISBN number of this book.

Many thanks to Mrs. Angela Rennwanz (University of Düsseldorf) for her help with the layout of this book!

Düsseldorf   Wilfried Linder

March 2016

Preface to the Third Edition

Also the second edition was sold successful. It seems that the hope I wrote about in chapter 6.8 ("A view into the future: Photogrammetry in 2020") will be fulfilled—photogrammetric techniques are not only in use until today but even new fields of applications came up. One of them is stereo photogrammetry with high resolution satellite images about which we will talk and learn in a new tutorial, see chapter 6.6. Another interesting new chapter (6.7) deals with simple flatbed scanners which you can use to create anaglyph images from small objects.

Again the software (included on the CD-ROM) was improved, a new programme (LISA FFSAT) was added, and the text in this book was actualised to the new possibilities of the software.

This is the place to thank the publisher and in particular Dr. Christian Witschel for the pleasant and straightforward collaboration since nearly 10 years!

January 2009   Wilfried Linder
Düsseldorf

Preface to the Second Edition

During the short time between the first edition and now many things happen giving the editors and me the idea not only to actualise this book but also to include further chapters. The changes are (among others):

The subtitle. It was the goal to give readers a compact and practical course with theoretical background only as far as necessary. Therefore we changed the subtitle from "Theory and Applications" to "A practical course". Nevertheless, and this was a remark of several reviewers, some more theory than before is included.

More about close-range photogrammetry. The first edition dealt mainly with aerial photogrammetry, now the field of terrestrial or close-range applications is expanded. For instance, an automatic handling of image sequences (time series) was developed and will be presented.

In this context we also take a special look to digital consumer cameras which now are available for low prices and which the reader may use for own projects in close-range applications. Regarding the lens distortion of such cameras, a chapter dealing with lens calibration was added.

A glossary now gives the reader a quick reference to the most important terms of photogrammetry. All words or technical terms included there are written in *italics* in this book.

Last but not least: The software which you find on the CD-ROM was improved and expanded, and the installation of software and data is now easier than before.

July 2005   Wilfried Linder
Bad Pyrmont

Preface to the First Edition

Photogrammetry is a science based technology with more than a century of history and development. During this time, the techniques used to get information about objects represented in photos have changed dramatically from pure optic-mechanical equipment to a fully digital workflow in our days. Parallel to this, the handling became easier, and so it's possible also for non-photogrammetrists to use these methods today.

This book is especially written for potential users which have no photogrammetric education but would like to use the powerful capabilities from time to time or in smaller projects: Geographers, Geologists, Cartographers, Forest Engineers who would like to come into the fascinating field of photogrammetry via "learning by doing". For this reason, this book is not a textbook—for more and deeper theory, there exists a lot of literature, and it is suggested to use some of this. A special recommendation should be given to the newest book from KONECNY (2002) for basic theory and the mathematical backgrounds or to the book from SCHENK (1999) for the particular situation in digital photogrammetry. For a quick reference especially to algorithms and technical terms see also the Photogrammetric Guide from ALBERTZ & WIGGENHAGEN (2005).

This book includes a CD-ROM which contains all you need from software and data to learn about the various methods from the beginning (scanning of the photos) to final products like ortho images or mosaics. Starting with some introductory chapters and a little bit of theory, you can go on step by step in several tutorials to get an idea how photogrammetry works. The software is not limited to the example data which we will use here—it offers you a small but powerful Digital Photogrammetric Workstation (DPW), and of course you may use it for your own projects.

Some words about the didactic principle used in this book. In Germany, we have an old and very famous movie, "Die Feuerzangenbowle" with Heinz Rühmann. This actor goes to school, and the teacher of physics explains a steam engine:

"Wat is en Dampfmaschin? Da stelle mer us janz dumm, un dann sage mer so: En Dampfmaschin, dat is ene jroße, schwachze Raum..." (SPOERL, 1933.

A language similar to German, spoken in the area of Cologne; in English: What is a steam engine? Suppose we have really no idea, and then let's say: A steam engine, that is a big black hole...). This "suppose we have no idea" will lead us through the book—therefore let's enter the big black hole called photogrammetry, let's look around and see what happens, just learning by doing. Theoretical background will only be given if it is indispensable for the understanding, but don't worry, it will be more than enough of theory for the beginning!

Concerning the object(s) of interest and the camera position(s), we distinguish between terrestrial (close-range) and aerial photogrammetry. This book mostly deals with the aerial case. Nevertheless, the mathematical and technical principles are similar in both cases, and we will see an example of close-range photogrammetry in the last tutorial.

A briefly description of the software is included in the last part of this book (chapter 10).

This is the right place to give thanks to all people who helped me:

To my chief, Prof. Dr. Ekkehard Jordan, for all the time he gave me to write this book, and for his interest in this science—he was one of the first Geographers using analytical photogrammetric methods in glacier investigation—and to all my friends and colleagues from the Geographic Institute, University of Düsseldorf, for many discussions and tests. To Mrs. Angela Rennwanz from the same institute—she made the final layout, therefore my special thanks to her!

To Prof. Dr. mult. Gottfried Konecny, who encouraged, helped and forced me many times and gave me a lot of ideas, and to all my friends and colleagues from the Institute of Photogrammetry and GeoInformation (IPI), University of Hannover, for their scientific help and patience—especially to my friend Dr.-Ing. Karsten Jacobsen. To Prof. Dr.-Ing. Christian Heipke, now chief of the IPI, who agreed that I could use all of the infrastructure in this institute, and for several very interesting discussions especially concerning image matching techniques.

For proof-reading of this book thanks (in alphabetical order) to Dr. Jörg Elbers, Glenn West and Prof. Dr. mult. Gottfried Konecny.

Un agradecimiento de corazón a mis amigos del America del Sur, especialmente en Bolivia y Colombia!

It may be of interest for you: All figures in this book are also stored on the CD-ROM (directory ...\figures) as MS PowerPoint™ files. Whenever you would like to use some of them, may be for education or scientific texts, please refer to this book! Thanks to the publishers for this agreement.

March 2003   Wilfried Linder
Bad Pyrmont

Contents

List of Figures

Formulas

Chapter 1
Introduction

1.1 Basic Idea and Main Task of Photogrammetry

If you want to measure the size of an object, let's say the length, width and height of a house, then normally you will carry this out directly at the object. Now imagine that the house didn't exist anymore—it was destroyed, but some historic photos exist. Then, if you can determine the scale of the photos, it must be possible to get the desired data.

Of course you can use photos to get information about objects. This kind of information is different: So, for example, you may receive *qualitative data* (the house seems to be old, the walls are coloured light yellow) from photo interpretation, or *quantitative data* like mentioned before (the house has a base size of 8 by 6 m) from photo measurement, or information in addition to your background knowledge (the house has elements of the "art nouveau" style, so may be constructed at the beginning of the 20th century), and so on.

Photogrammetry provides methods to give you information of the second type, quantitative data. As the term already indicates, photogrammetry can be defined as the "science of measuring in photos", and is traditional a part of geodesy, belonging to the field of remote sensing (RS). If you would like to determine distances, areas or anything else, the basic task is to get object (terrain) co-ordinates of any point in the photo from which you can then calculate geometric data or create maps.

Obviously, from a single photo (two-dimensional plane) you can only get two-dimensional co-ordinates. Therefore, if we need three-dimensional co-ordinates, we have to find a way how to get the third dimension. This is a good moment to remember the properties of human vision (see also Sect. 4.4). We are able to see objects in a spatial manner, and with this we are able to estimate the distance between an object and us. But how does it work? As you know, our brain at all times gets two slightly different images resulting from the different positions of the left respectively the right eye and according to the fact of the eye's central perspective.

© Springer-Verlag Berlin Heidelberg 2016
W. Linder, *Digital Photogrammetry*, DOI 10.1007/978-3-662-50463-5_1

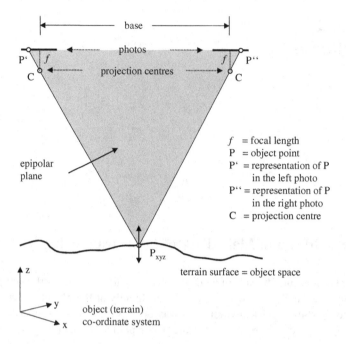

Fig. 1.1 Geometry in an oriented stereo model. Changing the height in point P (on the surface) leads to a linear motion (*left—right*) of the points P′ and P″ within the photos along *epipolar lines*

Exactly this principle, the so-called *stereoscopic viewing*, is used to get three-dimensional information in photogrammetry: If we have two (or more) photos from the same object but taken from different positions, we may easily calculate the three-dimensional co-ordinates of any point which is represented in both photos. Therefore we can define the main task of photogrammetry in the following way: For any object point represented in at least two photos we have to calculate the three-dimensional object (terrain) co-ordinates. This seems to be easy, but as you will see in the chapters of this book, it needs some work to reach this goal...

For the first figure, let's use the situation of aerial photogrammetry. To illustrate what we have said before, please take a look at Fig. 1.1.

Each point on the terrain surface (object point) is represented in at least two photos. If we know or if we are able to reconstruct all geometric parameters of the situation when taking the photos, then we can calculate the three-dimensional co-ordinates (x, y, z) of the point P by setting up the equations of the rays [P′ → P] and [P″ → P] and after that calculating their intersection. This is the main task of photogrammetry as you remember, and you can easily imagine that, *if* we have reached this, we are able to digitise points, lines and areas for map production or calculate distances, areas, volumes, slopes and much more.

1.2   Why Photogrammetry?

There are many situations in life or science in which we must measure co-ordinates, distances, areas or volumes. Normally we will use tools like a ruler or a foot rule. This is the place to discuss situations in which photogrammetric techniques may be used as an alternative or in which photogrammetry is the only possible way to measure:

In many cases the methods of measurement depend on the kind of the objects. As already mentioned in Sect. 1.1 it may happen that the object itself doesn't exist anymore but we have photos from the object. Similar to this are situations in which the object cannot be reached. For instance, imagine areas far away or in countries without adequate infrastructure, which then can be photographed from an aircraft to create maps.

Measure in photos means also measure without a physical contact to the object. Therefore, if you have very smooth objects like liquids, sand or clouds, photogrammetry will be the tool of choice.

Further, all kind of fast moving objects will be measured with photogrammetry. For instance these may be running or flying animals or waves. In industry, high-speed cameras with simultaneous activation are used to get data about deformation processes (like crash tests with cars).

In various applications, nowadays laser scanner equipment is an alternative to photogrammetry. In the aerial case laser scanning is used to get information about the relief (terrain models), but also in the close-range case these techniques are widely spread especially if it is necessary to get large amounts of three-dimensional point data (point clouds). The advantage here is that the object can be low textured —a situation where photogrammetric matching techniques (Sect. 4.6) often fail. On the other hand, laser scanning is time consuming and up to now expensive, comparing with photogrammetric methods, and laser scanning cannot be used for fast moving objects. Therefore, these methods may be seen as a supplement to photogrammetry.

1.3   Image Sources: Analogue and Digital Cameras

The development of photogrammetry is closely connected with that of aviation and photography. During more than 100 years, photos have been taken on glass plates or film material (negative or positive). In principle, specific photogrammetric cameras (also called *metric cameras*) work the same way as the amateur camera you might own. The differences result from the high quality demands which the first ones must fulfil.

Beside high precision optics and mechanics, aerial cameras use a large film format. You may know the size of 24 by 36 mm from your own camera—aerial cameras normally use a size of 230 by 230 mm (9 by 9 in)! This is necessary to receive a good ground resolution in the photos. As a result, the values of "wide angle", "normal" and "telephoto" focal lengths differ from those you may know—for example, the often used wide angle aerial camera has a focal length of about 153 mm, the normal one a focal length of about 305 mm.

Furthermore, the lens system of aerial cameras is constructed as a unit with the camera body. No lens change or "zoom" is possible to provide high stability and a good lens correction. The focal length is fixed, and the cameras have a central shutter.

Since some decades, manufacturers like Z/I imaging, Leica or Vexcel have been developing digital aerial cameras. As we can see today, there are two construction strategies. One is to keep the central perspective principle well-known from existing film cameras with the advantage that you can use existing software to handle the data. For this solution (called *frame camera*), a large area sensor is required. At the beginning, efforts are made to use four overlapping smaller sensors of industrial standard and then match the four image parts together, but nowadays large sensors are available and used (like by the DMC series in Fig. 1.2). The other strategy is to use a *line sensor* across the flight direction and collect data continually during the flight (ADS 80 from Leica). This is a bit similar to the techniques known from sensors on satellites or from hyper-spectral scanners.

For close-range applications there are several cameras of high quality on the market like those from Leica (X series, small-format) or Hasselblad (H5D series, medium-format with up to 60 megapixels, also well suitable for the aerial case, see Fig. 1.3).

Nowadays even digital consumer cameras have reached a high technical standard and good geometric resolution and are available for low prices. These cameras can be used in photogrammetry for applications with medium accuracy claim.

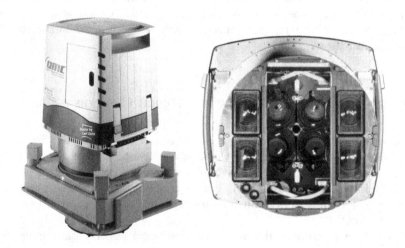

Fig. 1.2 The DMC (Digital Mapping Camera) from Z/I imaging—an example of a digital aerial camera. *Left* camera mounted on carrier. *Right* view from below—you can see the lenses belonging to the four area sensors. Courtesy of Intergraph Corp., USA

Fig. 1.3 Example of a semi-metric digital camera: The medium-format H5D from Hasselblad. Courtesy of Hasselblad Vertriebsgesellschaft m.b.H., Germany

1.4 Digital Consumer Cameras

As mentioned just before, various types of digital consumer cameras are on the market which may also be used for photogrammetric applications. The differences of the construction principles between metric and consumer cameras can be seen in general in quality and stability of the camera body and the lens. Further, consumer cameras usually have a "zoom" lens with larger distortions which are not constant but vary for instance with the focal length, so it is difficult to correct them with the help of a calibration.

If you want to purchase a consumer camera to use it for photogrammetry please take the following remarks into account:

General: It should be possible to set the parameters focal length, focus, exposure time and f-number manually, at least as an option.

Resolution (Number of pixels): The higher the number of pixels, the better—but not at any price: Small chips with a large number of pixels of course have a very small pixel size and are not very light sensitive, furthermore the signal-noise ratio is less good. This you will find especially with higher ISO values (200 and more) and in dark parts of the image. Therefore the pixel size or "pixel pitch" should not be less than 4 μm.

Distance setting (focus): It should be possible to de-activate the auto focus. If the camera has a macro option you can use it also for small objects.

Exposure time, f-number: The maximum f-number (lens opening) should not be less than 1:2.8, the exposure time should have a range of at least 1 … 1/1000 s.

Image formats: The digital images are stored in a customary format like JPEG or TIFF. Important: The image compression rate must be selectable or, even better, the compression can be switched off to minimise the loss of quality.

Others: Sometimes useful are a tripod thread, a remote release and an adaptor for an external flash. Two sets of accumulators, a battery charger, additional memory cards, if need be a card reader and a good tripod complete the equipment. A final remark: As everywhere in life, "cheap" is not always equal to "good"! Therefore you should better proof the quality than the price...

To work with image data from a digital camera you need some information like the focal length or the size of the pixels on the CCD chip. In the tutorials 3 and 4 you will see how to handle those images.

1.5   Short History of Photogrammetric Evaluation Methods

In general, three main phases of photogrammetry can be distinguished concerning the techniques of the equipment used for evaluation and the resulting workflow. The transition from one phase to the following took a time of about 20 years or even more.

In the Sect. 1.1 you saw that, if we want to get three-dimensional co-ordinates of an object point, we must reconstruct the rays belonging to this point from the terrain through the projection centres into the central perspective photos, a procedure which we call *reconstruction of the orientation* or briefly *orientation*. In the first decades of photogrammetry this was done in a pure optical-mechanical way. The large, complicated and expensive instruments for this could only be handled with a lot of experience which led to the profession of a photogrammetric operator. Not only the orientations of the photos but also any kind of the following work like measuring, mapping and so on was carried out mechanically. In later times, this phase was named the *Analogue Photogrammetry*.

With the upcoming of computers, the idea was to reconstruct the orientation no more analogue but algorithmic—via formulas with their parameters (coefficients) being calculated and stored in the computer. The equipment became significantly smaller, cheaper and easier to handle, and was supplied with linear and rotation impulse counters to register hardware co-ordinates, and with servo motors to provide the ability to position the photos directly by the computer. Nevertheless, the work still was done with real (analogue) photos and still needed a high precision mechanical and optical piece of equipment, the so-called *analytical plotter*. According to that, this phase was called *Analytical Photogrammetry*.

As everybody knows, in the last decades the power of computers rose at breath-taking speed. So, why not use digital photos and do the work directly with the computer? Even a simple PC nowadays has power and storage capacity enough to handle high-resolution digital photos. That is the phase now: *Digital Photogrammetry*, and that's what we want to explain with the help of this book, the included software and some examples. The only remaining analogue part in the chain of a total digital workflow sometimes are the photos themselves when taken with traditional cameras on film, but also this era will end soon.

For existing photos on film or paper, we will need a high-precision scanner as the only special hardware periphery. And due to the fact that around the world hundreds of "classical" aerial cameras are in use—instruments with a lifetime of decades—and digital cameras are expensive up to now, photo production on film with subsequent scanning may be a usual way for further years in several regions of the world. On the other hand we must recognise that a totally digital workflow has much advantages and is much faster, and no film development is necessary, a fact which significantly decreases the costs.

1.6   Geometric Principles 1: Camera Position, Focal Length

To explain the relation between the distance [camera position—object] (in the aerial case: flying height) and the focal length, we use a terrestrial example. First, take a look at Fig. 1.4.

Fig. 1.4 Different positions and lens angles. The situation, view from above

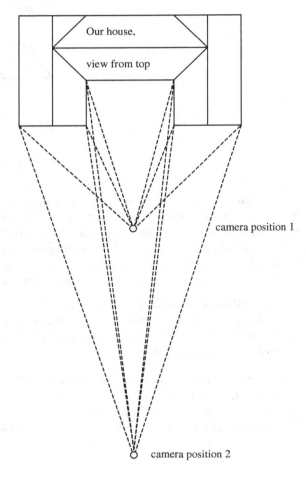

Our house,

view from top

camera position 1

camera position 2

Fig. 1.5 The results: Photos
showing the house in same
size but in different
representations due to the
central perspective

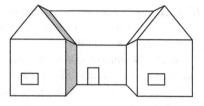

Photo taken from camera position 1

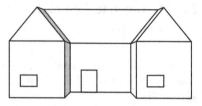

Photo taken from camera position 2

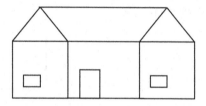

Photo taken from a position infinitely far away

Our goal is to take a photo of the house, filling the complete image area. We
have several possibilities to do that: We can take the photo from a short distance
with a wide-angle lens (like camera position 1 in the figure), or from a far distance
with a small-angle lens (telephoto, like camera position 2), or from any position in
between or outside. Obviously, each time we will get the same result. Really?

Figure 1.5 shows the differences. Let's summarise them:

- The smaller the distance [camera—object] and the wider the lens angle, the
  greater are the displacements due to the central perspective, or, vice versa:
- The greater the distance [camera—object] and the smaller the lens angle, the
  smaller are the displacements.

In a (theoretical) extreme case, if the camera could be as far as possible away
from the object and if the angle would be as small as possible ("super telephoto"),
the projection rays would be nearly parallel, and the displacements near to zero.
This is similar to the situation of images taken by a satellite orbiting some hundreds
of kilometres above ground, were we have nearly parallel projection rays but also

influences coming from the earth curvature. The opposite extreme case are photos taken with a *fisheye* lens which have an opening angle of up to 180°, some-times called whole-sky-systems.

What are the consequences? If we would like to transform a single aerial image to a given map projection, it would be the best to take the image from as high as possible to have the lowest displacements—a situation similar to satellite images (see above). On the other hand, the *radial-symmetric displacements* are a pre-requisite to view and measure image pairs stereoscopically as you will see in the following chapters, and therefore most of the aerial as well as terrestrial photos you will use in practise are taken with a wide-angle camera, showing relatively high relief-depending displacements.

1.7 Geometric Principles 2: Image Orientation

As already mentioned before, the first step of our work will be the reconstruction of the orientation of each photo, which means that we have to define the exact position of all photos which we want to use within the object (terrain) co-ordinate system. Now please imagine the following: If we know the co-ordinates of the projection centre, the three rotation angles (against the x-, y- and z-axis) as well as the focal length of the camera (part of the interior orientation, see Sect. 4.2.2), then the position of the photo is unequivocally defined (see Fig. 1.6). Therefore our first goal will be to get the six *parameters of the exterior orientation* (x_0, y_0, z_0, φ, ω, κ; see Sect. 4.2.5).

In the case of aerial photos, the values of φ (phi) and ω (omega) will normally be near to zero. If they are exactly zero, we have a so-called *nadir photo*. But in practice, this will never happen due to wind drift and small movements of the aircraft. Always remember the rule "nothing is exact in real life"! The value of κ (kappa) is defined as "east = zero" according to the x-axis of the terrain co-ordinate system, then counting anti-clockwise in grads, defining north = 100, west = 200, south = 300 grads (see Sect. 1.10 for the units).

Please note that only exact nadir photos of a true horizontal plane would have a unique scale or, in other words, non-zero values of φ and/or ω as well as the form of the object (for instance the relief) lead to scale variations within the photo.

If M_b is the mean photo scale or m_b the mean photo scale number, h_g the height of the projection centre above ground and f the focal length, we can use the following formulas (see Fig. 1.7):

$$m_b = h_g/f \qquad \text{or} \qquad M_b = 1/m_b = f/h_g$$

$$(1.7.1)$$

Fig. 1.6 Focal length, projection centre and rotation angles

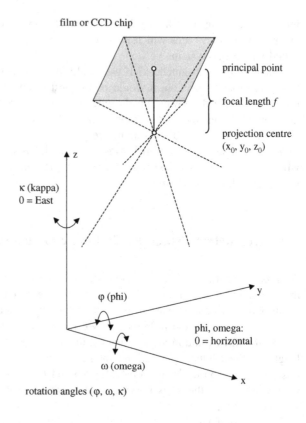

Fig. 1.7 Relations between focal length f, height above ground h_g and the photo scale f/h_g

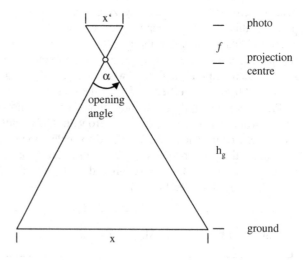

Now take a look at the different co-ordinate systems (CS) which we have to deal with. First, the camera itself has a two-dimensional CS; this may be a traditional or a digital one (*image CS*). Second, in case of film or paper material we must use a scanner which has a two-dimensional pixel matrix (*pixel CS*)—the equivalent to the photo carrier co-ordinates of an analytical plotter (see Sect. 1.5). And finally our results should be in a three-dimensional *object* (*terrain*) *CS*—normally a rectangle system like used for the Gauss-Krueger or the related UTM projection, connected with an ellipsoid to define the elevation (for instance, in Germany the Gauss-Krueger system is related with the Bessel 1841 or the WGS84 ellipsoid, the UTM system with the ellipsoid defined by Hayford 1924 or also the WGS84).

As we will see later on, the values of the three rotation angles depend on the sequence in which they were calculated. Often used are the sequences φ, ω, κ and ω, φ, κ—most software packages have the option to convert the angle values between these sequences.

1.8 Geometric Principles 3: Relative Camera Positions (Stereo)

To get three-dimensional co-ordinates of object points we need at least two images from our object, taken from different positions, as we already said in Sect. 1.1. This leads to the question which rules we must fulfil concerning the relative camera positions.

Remember Fig. 1.1: The point P(x, y, z) will be calculated as an intersection of the two rays [P′ → P] and [P″ → P]. You can easily imagine that the accuracy of the result depend among others from the angle between both rays. The smaller this angle, the less will be the accuracy: Take into account that every measurement of the image points P′ and P″ will have more or less small errors, and even very small errors here will lead to a large error especially in z when the angle is very small. Besides, this is a further reason why wide-angle cameras are preferred in photogrammetry (see next figure) (Fig. 1.8).

Let A be the distance between the cameras and the object and B be the distance between both cameras (or camera positions when only a single camera is used), then the angle between both projection rays (continuous lines) depend on the ratio A/B, in the aerial case called the *height-base ratio*. Obviously you can improve the accuracy of the calculated co-ordinates P(x, y, z) by increasing the distance B (also called *base*, see Fig. 1.1). If then the overlap area (stereo model, see next chapter) is too small you may use convergent camera positions—"squinting" in contrast to human vision (parallel). The disadvantage of this case is that you will get additional perspective distortions in the images. Please keep in mind: The parallel (aerial) case is good for human stereo viewing and automatic surface reconstruction, the convergent case often leads to a higher precision especially in z direction.

Fig. 1.8 Camera positions
parallel (*above*) and
convergent (*below*)

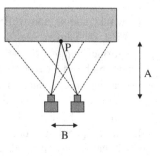

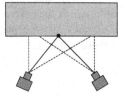

1.9   Some Definitions

Before starting with the practical work, we want to introduce some standard
technical terms of photogrammetry.

- *Photo*: The original photo on film
- *Image*: The photo in digital representation—the scanned film or the photo
  directly taken by a digital camera
- *Model* (*stereo model, image pair*): Two neighbouring images within a strip
- *Strip*: All overlapping images taken one after another within one flight line
- *Block*: All images of all strips
- *Base*: Distance between the projection centres of neighbouring images.

To illustrate what we mean, please take a look at the next figure (Fig. 1.9):

An image flight normally is carried out in the way that the area of interest is
photographed strip by strip, turning around the aircraft after every strip, so that the
strips are taken in a meander-like sequence. The two images of each model have a
longitudinal overlap of approximately 60–80 % (also called *endlap*), neighbouring
strips have a lateral overlap of normally about 30 % (also called *side lap*). As we
will see later on, this is not only necessary for stereoscopic viewing but also for the
connecting of all images of a block within an aerial triangulation (see Fig. 5.3).

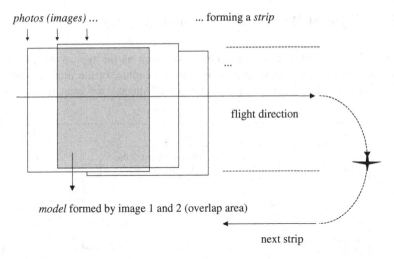

photos (images) ...   ... forming a *strip*

flight direction

model formed by image 1 and 2 (overlap area)

next strip

Fig. 1.9 Photos, models and strips forming a block

1.10   Length and Angle Units

Normally, for co-ordinates and distances in photogrammetry we use *metric* units, the international standard. But in several cases, also non-metric units can be found:

- Foot ('): Sometimes used to give the terrain height above mean sea level, for example in North American or British topographic maps, or the flying height above ground.
- Inch ("): For instance used to define the resolution of printers and scanners (dots per inch).

$$1' = 12" = 30.48 \text{ cm} \qquad 1" = 2.54 \text{ cm}$$
$$1\text{m} = 3.281' \qquad 1 \text{ cm} = 0.394"$$

(1.10.1)

You will surely know angles given in *degrees*. In mathematics also *radians* are common. In geodesy and photogrammetry, we use *grads*. In the army, the so-called *mils* are used.

A full circle has

$$360 \text{ degrees} = 400 \text{ grads} = 2\pi \text{ (pi)} = 6400^- \text{ (mils)}$$

(1.10.2)

1.11 A Typical Workflow in Photogrammetry

Finally let's take a look at the next figure, showing us the typical workflow for photogrammetric applications. Beginning with the capture of the images, we have then to calculate the orientation parameters of all images we want to use. After

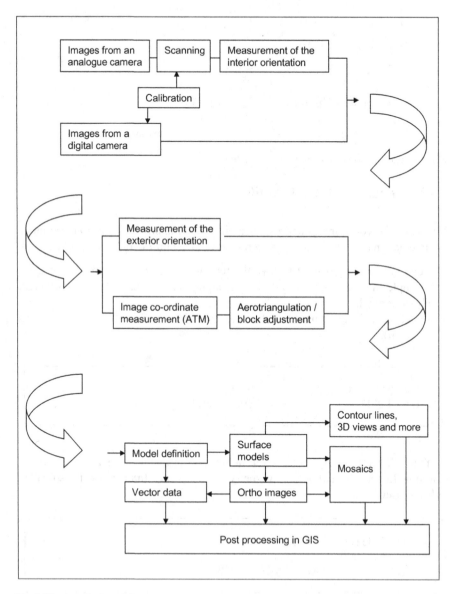

Fig. 1.10 A typical workflow

this we can measure co-ordinates, create several kind of image products like surface models, and also we may use the results in additional cartographic or GIS software (Fig. 1.10).

Chapter 2
Included Software and Data

2.1 Hardware Requirements, Operating System

If you want to use the software available from the Springer server and work with the example data or even use your own materials, it is necessary to have an adequate PC supplied with sufficient main memory (RAM), storage capacity (hard disk) and high resolution graphics. In particular, you need:

	Minimum	Recommended
Main memory (RAM)	1 GB	4 GB
Hard disk	10 GB	>> 50 GB
Graphics resolution	1024 x 768 pixels	1280 x 1024 pixels
Screen size	17"	21"
Mouse	3 buttons	central wheel

Furthermore, to handle (aerial) photos on paper or film material, you need a scanner (see Chap. 3). For stereoscopic viewing you need red-cyan glasses, a simple example is included in this book. You need a mouse with 3 buttons or with a central wheel which, when pressed down, also serves as middle mouse button.

For an ergonomic work you should use a "real" PC (not a tablet) with high-resolution graphics and a large screen, a keyboard and a mouse.

The software requires a MS Windows operating system, version 7 or higher (in 32- or 64-bits mode).

© Springer-Verlag Berlin Heidelberg 2016
W. Linder, *Digital Photogrammetry*, DOI 10.1007/978-3-662-50463-5_2

2.2   Image Material

In the first tutorials we will process aerial photos which were taken by an analogue aerial camera (see Sect. 1.3) in the usual format of 23 by 23 cm (9 by 9″) and then converted into a digital format using a scanner. Nevertheless, also images from non-metric, réseau or digital cameras, and not only aerial but also terrestrial photos can be handled.

From a practical point of view, for the following tutorials all image material is prepared on the Springer server. To help you handle your own examples, Chap. 3 will discuss the basic principles of scanning paper or film photos. Beside this, you may of course use images taken with a digital camera.

The aerial photos used in Chaps. 4 and 5 are owned by the Corporación Auto-noma del Valle del Cauca (CVC), Cali, Colombia. Thanks to Ing. Carlos Duque from the CVC who managed everything to give me the rights using these photos here.

The photos used in the Chaps. 6 and 7 are owned by the Institute of Photogrammetry and GeoInformation (IPI) of the University of Hannover, Germany. Thanks to Dr.-Ing. Folke Santel for her patience and help.

Section 9.1 deals with high resolution satellite images. For our tutorial we will use images from the Cartosat-1 satellite, showing an area south-west of Warszawa, Poland. Thanks to the Space Application Centre ISRO, Ahmedabad, India, and to GEOSYSTEMS Polska, Warszawa, for the courtesy to use the data (images and control points) in this book!

2.3   Overview of the Software

On the Springer server (extras.springer.com) you find a small but really useful digital photogrammetric software package with which you can make everything described in the following chapters and much more. In particular, the software is *not* limited to the example data but can be used for a wide range of photogrammetric tasks. The package is divided into four parts:

LISA BASIC: A raster GIS software with a lot of possibilities in image processing, terrain modelling and more. Copyright by the author.

LISA FOTO: Extension of LISA BASIC, digital photogrammetric workstation. This is the main software used in the following chapters. Copyright by the author.

LISA FFSAT: Digital photogrammetry for stereo satellite data. Developed by the author in cooperation with Dr.-Ing. Karsten Jacobsen, University of Hannover.

Please note: The LISA programmes delivered with this book are special versions with slightly reduced functionality and the maximum size per image is limited to 20 MB. A complete programme description will be copied onto your PC during the installation (see c:\program files (×86)\lisa\text\lisa.pdf).

BLUH: A professional bundle block adjustment software optimised for aerial triangulation. A "light" version including the central five modules of this programme system with reduced functionality and limited to a maximum of 30 images per block will be installed on your computer. Copyright by Dr.-Ing. Karsten Jacobsen from the Institute of Photogrammetry and GeoInformation, University of Hannover, Germany.

2.4   Installation

Important: Log into the operating system with full rights, usually select user = administrator. Copy the software from the Springer server to your PC and start the SETUP programme. You can select whether you like to create entries in the start menu and/or desktop icons. *For consistency with the example data it is urgently recommended to use the proposed program directory (installation path)!* Finally copy the data sets used in the following tutorials to your PC: Data of tutorial 1 to c:\program files (×86)\lisa\tutorial_1 and so on.

After the installation has finished, you will find the following additional directories on your PC:

c:\program files (x86)\lisa	LISA and BLUH programme files, fonts, runtime libraries etc.
c:\program files (x86)\lisa\text	the manual, PDF format
c:\users\public\lisa\pal	directory for palettes
c:\users\public\lisa\sig	directory for area symbols
c:\users\public\lisa\flt	directory for filter matrices
c:\users\public\lisa\cam	directory for cameras
c:\ program files (x86)\lisa\tutorial_1	data prepared for tutorial 1
...	...
c:\ program files (x86)\lisa\tutorial_5	data prepared for tutorial 5

Note: If your graphics has the required resolution (minimum 1024 × 768 pixels, 24-bits colour depth) but the error message "Screen resolution too small" appears, try the following: Click onto a LISA icon (for instance in the Start menu) with the right mouse button, then go to **Properties**, select the tab **Compatibility** and activate **Disable display scaling on high DPI settings** in the **Settings** section.

2.5   General Remarks

During the standard installation process, you have only those data files copied onto
your hard disk which are used as input files in the following tutorials (see Sect. 2.4).
Besides, many of the intermediate and final results are also prepared on the Springer
server (sub directory data\tutorial_x\output) and can be used for control purposes
or, if you would like to skip some steps and go on later, to get intermediate results
necessary for the following steps. Therefore, at the end of any tutorial chapter all
created files are listed.

For consistency it is recommended to use the file names proposed in the tutorials. In general, it is of course possible to choose any output name.

To make the work a bit clearer in the following tutorials, special fonts are used:

- **Options** and **parameters**: For instance, **Image No.** refers to the corresponding
  text in an input window.
- Menu entries: Separated by ">", for example: **Processing > Stereo mea-
  surement** means that you first have to click onto **Processing**, then onto **Stereo
  measurement**.
- *Definitions* or *key words* are printed in italics.
- Any results stored in a file and listed here for control purposes are printed in
  this font.
- File names are always printed in UPPERCASE letters.
- Units are printed in [square brackets], example: [μm].
- Vectors are also printed in square brackets with an arrow showing the direction
  like [start point → ending point].

See also Sect. 10.4 for some remarks about the programme handling.

Chapter 3
Scanning of Photos

3.1 Scanner Types

A lot of scanners exist on the market with differences in construction, geometric and radiometric resolution, format size and last but not least price. For use in photogrammetry, some basic requirements must be fulfilled: Format A3, transparency unit (for film material), high geometric and radiometric resolution and accuracy.

The format A3 is necessary because for photogrammetric purposes the photos must be scanned in total, in particular including the fiducial marks (see Sect. 4.2.1), and most of the aerial photos usual today have the format 23 by 23 cm (9 by 9″) which exceeds the A4 format. On the other hand, the side information bar (mostly black; contains additional information like altimeter, clock, film counter) should not be scanned to save storage capacity.

In low-cost photogrammetry often flatbed (DTP) scanners are used with a geometrical accuracy of about 50 μm (see for instance Wiggenhagen (2001)). For a better understanding, three important aspects of influence shall be mentioned:

- Accuracy along the CCD array (charge coupled device; under the moving bridge beneath the glass plate): Constancy of size, distance and linear arrangement of the CCD elements.
- Accuracy across the CCD array (in moving direction of the bridge): Constancy of step width and linearity of the moving.
- Angle between bridge and moving direction: Deviations from a rectangle.

Some words about the radiometric resolution: The absolute minimum a photogrammetric scanner must have is the possibility to scan grey scale (panchromatic) photos with 8 bits which means 256 grey levels. In case of colour photos, normally we need a radiometric resolution of 24 bits which means 8 bits or 256 levels for each of the three base colours (red, green, blue), scanned in single-pass mode.

© Springer-Verlag Berlin Heidelberg 2016
W. Linder, *Digital Photogrammetry*, DOI 10.1007/978-3-662-50463-5_3

3.2   Geometric Resolution

The geometrical scan resolution is given in the units "dots per inch" [dpi] or micrometres [µm] and reflects on the maximum accuracy to attain. For simple photogrammetric investigations as shown in this book, a value of 300 or 600 dpi may be used. A scan resolution of 600 dpi (42 µm) is near to the geometric accuracy of most flatbed scanners (about 50 µm, see above). The conversion from [dpi] to [µm] is based on the formula:

$$\text{pixel size in [µm]} = 25400 \,/\, \text{resolution in [dpi]}$$
$$\text{resolution in [dpi]} = 25400 \,/\, \text{pixel size in [µm]}$$

(3.2.1)

The table below serves to illustrate the relation between scan resolution in [dpi] or [µm], the image size in [MB] (grey scale/8-bit photo), the aerial photo scale and the ensuing pixel size in terrain units, usually [m]:

Resolution [dpi]	150	300	600	1200	2400	4800
Pixel size [µm]	169.3	84.7	42.3	21.2	10.6	5.29
Image size ca. [MB]	2	8	32	128	512	2018
Photo scale						
1: 5000	0.847	0.423	0.212	0.106	0.053	0.026
1: 7500	1.270	0.635	0.318	0.159	0.079	0.040
1:10000	1.693	0.847	0.423	0.212	0.106	0.053
1:12500	2.117	1.058	0.529	0.265	0.133	0.066
1:15000	2.540	1.270	0.635	0.317	0.159	0.079
1:17500	2.963	1.482	0.741	0.370	0.175	0.093
1:20000	3.386	1.693	0.846	0.424	0.212	0.106
1:25000	4.233	2.117	1.058	0.529	0.265	0.132
1:30000	5.080	2.540	1.270	0.634	0.318	0.159
1:40000	6.772	3.386	1.693	0.846	0.424	0.212
1:50000	8.466	4.234	2.116	1.059	0.530	0.265

Pixel size in terrain units ca. [m]

For the geometrical scan resolution it is a good idea always to follow the rule "As high as necessary, as low as possible"! On the other hand, the maximum attainable accuracy in z (altitude) depends, among other factors, on the scan resolution. The accuracy in z can reach a value of 0.1 ‰ (per thousand) of the flying height above terrain, using an analytical plotter and photos with an endlap of 60 %.

3.3 Some Practical Advice

It is suggested that all photos are first arranged on a table in the same position and orientation in which they form the block. This means that, for example, all photos are situated with "top = north" independent from the position of the side information bar. Then every photo is scanned "westeast parallel to the CCD array". This method has the advantage that the resulting digital images are arranged in the same way as they follow in the strip.

- If at all possible only master film material should be used as scan sources. If film is not available prints must be used instead. They should be processed on plain (non-textured) paper of high geometrical stability.
- Please note that the *whole* aerial photo must be scanned—in particular, the fiducial marks must be included, which we will need to establish the *interior orientation* (see Sect. 4.2.2). On the other hand, the photo borders and the side information bar should not be scanned to save memory space.
- Grey scale images should be stored as "grey scale", not as "colour" images! The standard file formats to choose for storing and later to import into LISA FOTO are BMP, JPEG, PNG or TIFF (uncompressed). Please note: The file extensions used by LISA are JPG (not JPEG) or TIF (not TIFF)!
- Image names: As a general rule, the image *names* should be identical with the image *numbers* with no other or further text. Example: Image No. 137 will be stored, depending on the format, as 137.BMP, 137.JPG or 137.TIF, but *not* as LEFT.BMP, FOTO_137.BMP or anything else.
- Some general remarks for scanning: Switch on the scanner without a photo on the glass plate! Let the equipment run at least 5 minutes to warm up. After that, put the photo onto the glass plate and cover the unused area of the plate with a black cardboard. In this way, the radiometric self-calibration of the scanner is supported.

3.4 Import of the Scanned Images

LISA uses a special image format with the extension IMA. In the appendix you find information about the file formats. Within the programme, you can directly use also image data in one of the formats BMP, JPG, PNG or TIF. Nevertheless, it may be more comfortable to have all data in the same format, therefore let's take a look onto the import option.

Please start LISA FOTO, then select the option File > Import. Choose the format (BMP, JPG, PNG or TIF). As you will see, you have two additional options used simultaneously for all images which shall be explained here:

- **Rotate by 0, 90, 180, 270 degrees:** For instance, chose 180, if the photos where scanned against our general rule concerning the position on the glass plate.
- **Connect images to camera:** Special option for images sequences, see also Chap. 7.

Now click either onto the All or the **Selection** button—in the latter case, use the ctrl-key and the left mouse button to mark the images you want to import. A protocol window appears showing each imported image. After the last file is processed, the window will be closed.

Beside of this, you have of course the usual "**save as**" option in the image display of LISA to convert an image between the formats mentioned above: Load for instance a BMP file and save it as PNG.

Chapter 4
A Single Model

4.1 Project Definition

To work with the LISA programmes, it is necessary first to define a *project* or to select one already existing. All projects which we will use during the following tutorials are prepared and have been copied onto your computer in the installation process. Nevertheless, this is a good moment to take a look at this topic.

Start LISA FOTO. In the first appearing window you will be asked if you want to:

- Use the last project
- Select an existing project or
- Define a new project

Please select the project TUTOR_1.PRJ, then click onto the OK button. Now go to File > Edit project. In the appearing window you will see some entries—let's talk about their meaning:

Name: This is also the name of the project definition file which has the extension PRJ and is at all times located in the LISA main directory, usually c:\program files (×86)\lisa.

Working directory: All data we need will be searched by the programme in this directory (folder, path). In the same way, all data which we create will be stored in this directory. The button beside the input field can be used to open a directory tree view useful to browse to the desired path. If you key in a directory which did not yet exit it will be created. Important: All projects used here are prepared for the drive (hard disk) C. If you use a different drive against our advice, let's say D, you have to correct the path in all PRJ files before starting LISA!

Image data base: Optional for the handling of geocoded images in large projects. We will not need this option in our tutorials, therefore it is not necessary to define a data base.

© Springer-Verlag Berlin Heidelberg 2016
W. Linder, *Digital Photogrammetry*, DOI 10.1007/978-3-662-50463-5_4

Furthermore, a project is defined by a co-ordinate range in x, y and z and a pixel size (=geometric resolution, in terrain units). The border values of x and y usually should be multiples of the pixel size. In particular:

The co-ordinate range in x and y should be set to the outer boundaries of the whole project area. In special cases they can be set to extremely large values using the Reset button—don't do this here! Then, they will not be taken into account.

The z value range is of importance wherever digital terrain models (DTMs) will be created. Because of the fact that DTMs are 16-bit raster images with a defined relation between the pixel grey values and the corresponding heights, it is necessary to fix this relation within the project, for example when single DTMs shall be matched or mosaicked.

To help to find the border co-ordinates of x, y and z, a reference file (geocoded image or vector file) may be used. For this, the buttons Ref. raster and Ref. vector are prepared.

The length units (terrain units) can be selected: μm, mm, m or km. For this example all values are given in meters.

Remark: The values of pixel size, minimum and maximum height are fixed for all data within one project! Therefore it is really necessary to set values that make sense for these parameters.

In our first example, we will use the following values (all in [m]):

```
X from 1137300 to 1140000
Y from 969900 to 971700
Z from 1000 to 1700
Pixel size (object) 5
```

Length unit: Select m [meters].

If you want to create a new project, you can use the described option when the programme starts, or use File > Define project from within the programme. In our case, just close the window, for example with the Esc key, or, if you have changed something (for instance, the path), click onto OK.

Created file: TUTOR_1.PRJ.

4.2 Orientation of the Images

4.2.1 Camera Definition

If we have image material coming from a film camera and then was scanned, the first step to orient an image is the so-called *interior orientation* which means establishing the relation between (1) the camera-internal co-ordinate system and (2) the pixel co-ordinate system (see Sect. 1.7). The first one is given by the so-called *fiducial marks* superimposed in the image and their nominal co-ordinates, usually given in [mm] in the *camera calibration certificate*. In this document you

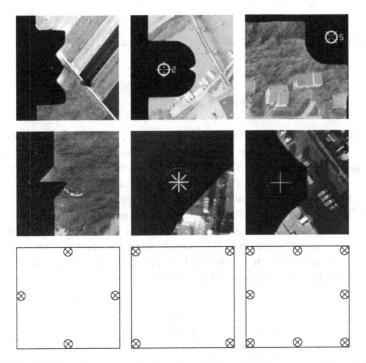

Fig. 4.1 Shapes (*first* and *second row*) and positions (*third row*) of fiducial marks in aerial photos

will also find the calibrated focal length in [mm]. After measuring (digitising) the marks, the software will be able to calculate the transformation coefficients for the relation between both systems (Fig. 4.1).

Some information about the fiducial marks: Older cameras have only 4 marks, situated either in the middle of the image borders (e.g. cameras from the Zeiss company, RMK series) or in the corners (e.g. cameras from the Wild company, RC series). Newer cameras have 8 marks, situated both in the middle of the borders and in the corners.

For the camera definition we need the nominal co-ordinates of the fiducial marks and the focal length, all given in [mm]. Usually we can get these data from the camera calibration certificate, see above. If this is not available, we can get the focal length from the side information bar of the images or, if even this is not possible, we can set the focal length to standard values. In case of analogue aerial cameras these are 153 mm (wide angle) or 305 mm (normal angle). For the fiducial marks we will then also use standard values (available for Zeiss RMK and Wild RC via the respective buttons).

In our first example we will use the following standard (nominal) data:

Fiducial No.	x value	y value
1	113.000	0.000
2	0.000	-113.000
3	-113.000	0.000
4	0.000	113.000

The focal length is 152.910 mm. In this example, we have images taken by a Zeiss RMK A 15/23 camera with 4 marks. Please start the option **Pre-programmes > Camera definition > Analogue**. In the appearing window, key in the values of x and y for each of the 4 fiducial marks, set the focal length and after that the name of the output file—in our case, please take RMK_1523.CMR. Or, just click onto the **Open** file button and load the prepared file into the window.

After clicking the **OK** button, the camera definition file will be created in the directory c:\users\public\lisa\cmr. For control purposes, start the text editor by clicking onto the **Text** button right-hand in the main window and open the file RMK_1523.CMR. The content must be like this:

```
    1    113.000       0.000
    2      0.000    -113.000
    3   -113.000       0.000
    4      0.000    113.000
  152.910
DP    -0.9999990000E+06      0.0000000000E+00
DP     0.0000000000E+00      0.0000000000E+00
PP     0.0000000000E+00      0.0000000000E+00
CS    5.000    5.000  159.806
```

Remarks: The camera definition must only be done once and is valid for all images taken with the same camera. The last four lines in the file are without meaning here.

Created file: RMK_1523.CMR

4.2.2 Interior Orientation

As mentioned before, the next step will be measuring (digitising) the fiducial marks to set up the transformation between camera and pixel co-ordinates. This must be done once for each image which you would like to use in further work.

Please start the option **Pre programmes > Orientation measurement** and select the file 157.IMA in the file manager. After loading this image, go to **Measure > Interior orientation**. The next window will ask you for the camera

Left: Fiducial mark of a   Right: Result of the automatic
Zeiss RMK camera (white,   centring (strongly zoomed)
point-shaped)

Fig. 4.2   Result of automatic centring of a fiducial mark

definition file—please use the just created one, RMK_1523.CMR. An option
Center is offered:

If the fiducial marks have the form of a white dot within a dark background, it is
possible to let the programme make an automatic centring onto the marks with
subpixel accuracy. In our example we have such marks, so please activate this
option (see Fig. 4.2).

Now click onto the **OK** button. The programme will automatically move the
image near to the first fiducial mark. Please note the measurement principle: Fixed
measuring mark, moving image like in analytical plotters! So, if you keep the
middle mouse button pressed down and move the mouse, the image will move
simultaneously "under" the measuring mark (default: red cross). Now move the
image until the first fiducial mark lies exactly under the measuring mark, then click
onto the left mouse button. If nothing happens, just move the mouse a little.

In this and all other display modules, you may vary the brightness and the
contrast to get a better impression of the image(s).

In the listing below you can see the measured co-ordinates, marked with M, and
in the main window you can recognise that the programme has moved the image
near to the second fiducial mark. Again move the image until fiducial and measuring mark are in the same position and click onto the left mouse button. In the
same way measure the third fiducial mark. And now a first test of accuracy: The
pre-positioning of the fourth fiducial mark should be very good, the displacement
should not exceed a few pixels. What is the reason?

After three fiducial marks are measured, the programme starts calculating the
transformation parameters (plane affine transformation, see below). If both the
nominal values from the camera definition and the measurements were exact
enough, the pre-positioning should be quite good. And an additional remark: This
calculation is done after each measured mark beginning with the third one. So, if we

would have 8 fiducial marks, then the pre-positioning should be better and better until the last mark is reached.

Examples for 2-dimensional transformations:

$x' = a_0 + a_1x + a_2y$ plane affine transformation
$y' = b_0 + b_1x + b_2y$ ... at least 3 points required

$x' = a_0 + a_1x + a_2y + a_3x^2 + a_4xy + a_5y^2$ 2nd order polynomial
$y' = b_0 + b_1x + b_2y + b_3x^2 + b_4xy + b_5y^2$ ... at least 6 points required

with (x, y) given co-ordinates, (x', y') new co-ordinates and
$a_0, a_1, ..., b_0, ...$ coefficients of the equation system

---

$$(4.2.1)$$

Back to our example: Measure the last (fourth) mark, move the mouse a little, and see the listing below which should be more or less like the following:

No.	x [mm]	y [mm]	Res. x	Res. y
1	113.000	0.000	0.014	−0.015
2	0.000	−113.000	−0.014	0.015
3	−113.000	0.000	0.014	−0.015
4	0.000	113.000	−0.014	0.015
Standard deviation [mm] :			0.014	0.015

Let's take a short look at the *residuals* (remaining errors after the adjustment): You can see that all of them have the same absolute values in x as well as in y or in other words, they are symmetrical. The reason is that with 4 points we have only a small over-determination for the plane affine transformation (at least 3 points are necessary). If you carry out an interior orientation of an image with 8 fiducial marks, the residuals will vary.

Now click onto the **Ready** button (checkmark). The programme will inform you about the calculated scan resolution in [dpi] and [μm]—in our case about 300 dpi or 84.7 μm. By this, you have a further check if the interior orientation was successful. Click onto **OK**, then close the window, for example with the **Esc** key.

For training purposes, please repeat what we have done here with our second image, 158. For this you may use the **open next image** button.

Created files: 157.INN, 158.INN.

Before going on, this is a good moment to talk about two different ways to complete the orientation. From the era of analytic photogrammetry you may know the three steps *interior—relative—absolute orientation*. Within the relative orientation the two images are "connected" by the calculation of model co-ordinates. Then, these are transformed to terrain co-ordinates in the absolute orientation.

In the following chapters we will take a different way: For each image we will first carry out the exterior orientation independently, may it be manually by measuring control points (see Sect. 4.2.5) or automatically, using a method called aerial triangulation (see Sect. 5.1). After that, neighbouring images are "connected" to form a model in a *model definition* (see Sect. 4.3 for instance).

4.2.3   Brightness and Contrast

It is one of the advantages of digital stereo photogrammetry that you can easily improve brightness and contrast "on the fly" when measuring within images. This is sometimes called a *photo lab at your fingertips*. Now, what happens?

The grey values of an image (range 0–255) are displayed with exactly these values used to set the brightness of each pixel. But, establishing a linear equation f (g) = c * g + b (c = contrast, g = grey value, b = brightness) between image and display, brightness and contrast can easily be changed. Figure 4.3 shows the results.

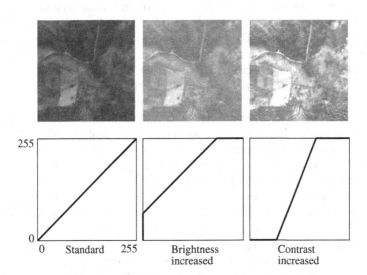

Fig. 4.3 Relations between grey values in the image and on the screen

4.2.4   Control Points

As described before, the final step within the orientation process will be calculating the relation between image and object co-ordinates, the so-called *exterior orientation*. For this we will have to measure ground control points, as you will see in the next chapter.

A *ground control point* (GCP) is an object point which is represented in the image and from which the three-dimensional object (terrain) co-ordinates (x, y, z) are known. In our case of aerial photogrammetry, this means that we have to look for points in our image, find these points for instance in a topographic map and get their co-ordinates out of the map, x and y by manual measurement, z by interpolating the elevation between neighbouring contours.

For each image we need at least 3 well-distributed GCPs. A basic rule is "the more, the better" to get a stable over-determination, therefore we shall look for at least 5 points (see also Sect. 4.2.6). "Well-distributed" means that a minimum of 3 points should form a triangle, not a line. Furthermore, best accuracy will be achieved in areas surrounded by GCPs. Last but not least it is not necessary but a good idea to use as many identical points as possible in neighbouring images forming a model later.

We can distinguish two kinds of GCPs, called *signalised (targeted)* and *natural* points. Often, before taking the photos, topographic points are signalised on the ground by white bars (size e.g. 1.2 by 0.2 m) forming a cross with the point itself marked with a central "dot" of e.g. 0.2 m diameter (all dimensions depending of course on the photo scale). The corresponding terrain co-ordinates are available from the Land Surveying Office or sometimes from the company taking the photos.

But often we have no signalised GCPs. Then we must look for real object (terrain) points which we can clearly identify in the image as well as in a topographic map mentioned before. But not every point is really good to serve as a GCP: As far as possible, choose rectangle corners (e.g. from buildings) or small circle-shaped points. These have the advantage to be scale-invariant. Take into account that we need also the elevation—this might be a problem using a point on the roof of a building, because it is not possible to get its elevation from the map! Therefore, if possible, prefer points on the ground.

Please remember that points may "move" during time, e.g. when lying on the shore of a river or at the border of a non-paved road. And also remember that the corresponding GCP position in a topographic map may be displaced as a result of map generalisation. Some idea which are good or poor points is shown in Fig. 4.4.

A powerful alternative to getting co-ordinates from object points then serving as GCPs is to use GPS equipment (*Global Positioning System*). The advantage is that you can use nearly every terrain point represented in the images and that you have no problems due to map generalisation. The disadvantage is of course that you have to go to your area and, to get really good results, carry out differential measurements (DGPS) with one receiver on a topographic point (*base*) and a second one used in the field at the same time (*field* or *rover*). The problems of "moving" points

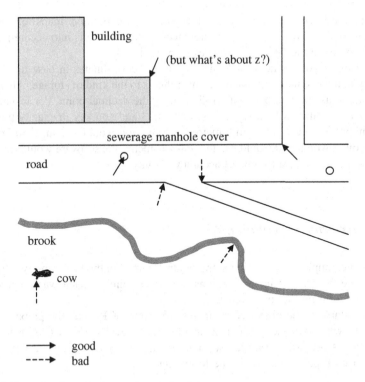

building

(but what's about z?)

sewerage manhole cover

road

brook

cow

→ good
----→ bad

Fig. 4.4 Examples for natural ground control points

mentioned before also may occur, the greater the time difference between the dates of taking the images and your GPS campaign. This is not the place to discuss GPS measurements—please use appropriate literature and the equipment's manual if you want to use this technique.

Aerial image flights nowadays are operated with simultaneously registration of the camera position (x_0, y_0, z_0 via DGPS) and the rotation angles (IMU, *Inertial Measurement Unit*), giving approximation values for the exterior orientation (see next chapter). We just have to convert the angles roll, pitch and yaw [degrees] into the photogrammetric system $\varphi - \omega - \kappa$.

To prepare the input of GCPs you can use the option **Pre programmes > Control points**. You can open an existing file or create a new one. Using the respective buttons you may edit, add or delete points. With the button **Ready** you will close and store the file. In this way you can handle a maximum of 900 control points per image.

There are two important aspects concerning the GCP terrain co-ordinates:

- In geodesy, the x axis shows to the north, the y axis to the east in a right-hand system. In photogrammetry, we use a mathematical co-ordinate system definition with x to the east, y to the north in a left-hand system. Whenever you get co-ordinates in form of a listing, labelled "x" and "y", make sure that this refers to the photogrammetric order! Furthermore, topographic maps of several countries also show the geodetic reference—please take this into account if you want to define the GCP co-ordinates from such maps.

- At all times you create a GCP file, key in the co-ordinates in base units, normally in meters, not in kilometres! This reflects to the kind of storage: All values are stored as real numbers with 3 digits after the decimal point. For instance, if you have a value of let's say x = 3250782.023 and you key in exactly this, the (nominal) accuracy is one millimetre. Imagine you would key in 3250.782023 or in other words you would use the unit kilometre, the software would only use 3250.782 meaning a (nominal) accuracy of only 1 m.

4.2.5 Exterior Orientation

In our first example, we will use natural control points. In the following two figures you can see their approximate positions. For each point a sketch was prepared as you will see during the measurement.

Before starting, the object co-ordinates from our GCPs must be prepared in a file, each with No., x, y and z in a simple ASCII format. Go to **Pre programmes > Control points** (see above). In the next window use the **Add** button for each point and key in the following values:

Point No.	X	Y	Z
15601	1137768.212	969477.156	1211.718
15602	1138541.117	969309.217	1245.574
15603	1139550.021	969249.250	1334.405
15701	1137534.649	970320.150	1251.964
15702	1138573.149	970388.650	1171.448
15703	1139623.149	970359.457	1158.972
15801	1137848.958	971643.004	1142.964
15802	1138601.712	971220.373	1157.148
15803	1139761.651	971315.870	1130.292
15901	1137598.525	972308.940	1128.694
15902	1138667.551	972228.208	1141.743
15903	1139767.051	972325.708	1144.467

After the last point is entered, click on the **Ready** button and store the file as CONTROL.DAT, then close the window. Or simply use the prepared file from the Springer server (…\tutorial_1\output).

Like for the interior orientation, the exterior orientation must be carried out once for each image. For the first image (No. 157) we will do this together step by step—for the second image (No. 158), again you will do this alone for training purposes.

If you have problems with this, as all times you may use the prepared results from
the Springer server (…\tutorial_1\output).

Please start **Pre programmes > Orientation measurement** and key in 157 as
name of the input image. After loading this file, go to **Measure > Exterior orientation.** The next window will ask you for the control point file; use the file
CONTROL.DAT just created before. The button **Reset** can (and should here) be
used to reset the projection centre co-ordinates to "unknown" (−999999) if they are
already existing. The option **Create point sketches** should be de-activated.

Now it is your turn to digitise the control points: You can select a point in the list
below—in our case, simply start with the first one, No. 15601. Use Fig. 4.5 to find
the approximate position of the point. In the small sketch window on the bottom left
side of the screen you will see the neighbourhood of the GCP which may help you
to find its exact position (this is the already prepared point sketch, see above—a
nice tool, but not necessary!). Move the image in the main window with the mouse,
middle mouse button depressed, until the GCP lies exactly "under" the measuring
mark—you will remember this process from the interior orientation (Sect. 4.2.2).
Now click onto the left mouse button. The point and its number will be superimposed in the image and marked with M in the list below. In the same way go on

Fig. 4.5 Positions of the control points in the left image (No. 157)

point by point until the last one for this image (No. 15803) is measured. If necessary, use the slider at right in the list window to scroll up or down.

After the fourth point is measured, a so-called *resection in space* from object to pixel co-ordinates is calculated by setting up the collinearity equations (see Sect. 4.3). As a consequence, for each further point a pre-positioning will be done by the programme and residuals as well as the standard deviation are calculated and shown in the list window.

When the last point was measured, click onto the **Ready** button (checkmark). The results in the list window may be like the following:

```
No.         x [mm]      y [mm]      Res. x      Res. y

15601      -67.027      71.969       0.024      -0.041 M
15602      -81.847      10.645      -0.063      -0.021 M
15603      -90.674     -73.878      -0.069       0.020 M
15701        0.166      91.835       0.009      -0.047 M
15702        5.048       7.430       0.002      -0.041 M
15703        2.418     -72.617      -0.014       0.053 M
15801      100.316      61.913       0.043      -0.007 M
15802       68.421       4.983       0.048       0.011 M
15803       73.802     -81.715       0.030       0.079 M

Standard deviation [mm]:             0.040       0.042
```

The residuals in x and y at every point as well as the resulting standard deviation are given in [mm] referring to the image. Remember the scan resolution of 300 dpi = 84.7 μm to see that the residuals are about half a pixel. In contrary to the results of the interior orientation (Sect. 4.2.2) you can see that the residuals are no more symmetric. We have measured 9 well-distributed points, much more than the minimum (3 points), and therefore a good over-determination is achieved (see next chapter). Now close the measurement window.

The results are stored in the file 157.ABS, let's look at the first three lines:

```
     152.910                           focal length [mm]
 -0.011145    0.011771    1.574385     rotation angles [radians]
1138648.415 970300.077 3152.826        projection centre [m]
```

The second and third line shows the six parameters of the exterior orientation, see Sect. 1.7.

For training purposes, please repeat the procedure of this chapter with the second image, 158. Use Fig. 4.6 to find the approximate positions of the GCPs. In this image start with point No. 15701. Again, point sketches are already prepared to help you to find the exact position.

Created files: CONTROL.DAT, 157.ABS, 158.ABS

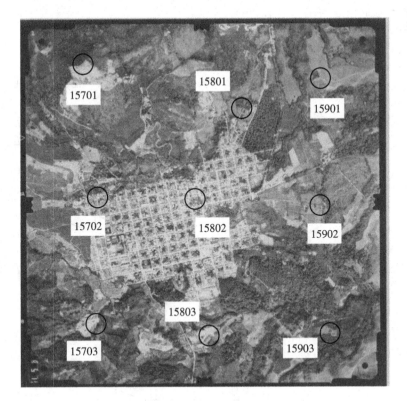

Fig. 4.6 Positions of the control points in the right image (No. 158)

4.2.6 Over-Determination and Error Detection

With Fig. 4.7 we want to explain the principles of over-determination. Let's imagine that we want to determine the parameters of a one-dimensional linear function, the general form given by $f(x) = ax + b$, by measuring values $f(x)$ at two or more positions of x. Mathematically such a function can unequivocally be fixed by only two points (observations). Part (a) shows this, but you can also see that wrong observations lead to a bad result. Part (b) illustrates an over-determination by three measured points, and as a result, the parameters of the function can be calculated using a least squares adjustment, furthermore the residuals r can be calculated for every point. These give us an idea about the quality of the observations, but in most cases we cannot decide what point is really bad because the residuals vary not very much. The result is better but not good. Part (c) shows the solution: With seven points we have a very good over-determination, and now it is clear to see that the central observation is wrong (a so-called *peak*). Deleting this, the adjustment gives us a good result.

Fig. 4.7 Calculated versus correct graph of the function f(x) = ax + b using two, three or more observations (r = residuals)

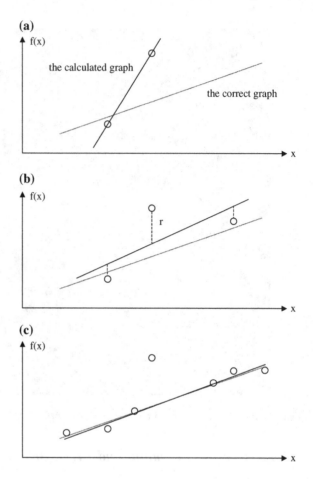

(a)

f(x)

the calculated graph

the correct graph

x

(b)

f(x)

r

x

(c)

f(x)

x

4.3   Model Definition

It is our goal to measure three-dimensional object co-ordinates, as you will remember from Sect. 1.1. Therefore all of the following steps will be done simultaneously in two neighbouring images, called the *stereo pair* or the *model* (see Sect. 1.9 and Fig. 1.9).

Before going on, the particular (actual) model must be defined. Please start Pre programmes > Define model. In the next window please set/control the following parameters:

Left image 157, right image 158, **parallax correction** 1 pxl. Exterior orientation: **Parameters** from ABS files (see Sect. 4.2.5), **Orientations** keeps empty, **Object co-ordinates** CONTROL.DAT.

After a short time, a property sheet window with several information will appear. Before explaining what happened, let's check the data:

Section **Range**: This is the model area (ranges of x, y and z) calculated by the programme. The values should be more or less like the following:

```
X from    1137185 to 1140038
Y from     969875 to  971776
Z from       1000 to    1700        (all values in [m])
```

In fact, the elevation range is *not* calculated but taken from the project definition as well as the pixel size (5 m, not displayed here; these values are fixed within the project, see Sect. 4.1).

Section **Geometry**: Pixel size in the digital image, resulting from photo scale and scan resolution, about 1.11 m. Ratio distance/base about 1.99. Maximum attainable accuracy in z, resulting from both values before, about 2.21 m. Photo scale about 1:13229. Please note that all values listed here are only given for control purposes and may differ a bit from those on your computer!

May be you remember the (ideal) value "0.1 ‰ of height above ground" (Sect. 3.2): The mean flying height is about 3100 m, the mean terrain height is about 1100 m, therefore we have more or less a value of 1 ‰. To save memory space, our photos were scanned with 300 dpi—if we would have scanned them with 600 dpi, the maximum attainable accuracy in z would be about 0.5 ‰ in this example.

Now click onto **OK** again—the model is defined. But not only this: In between, a lot of calculations and logic tests were carried out, and this is a good place to explain some of them in addition with a bit of theory.

As you will remember from Sect. 1.1, 1.7 and Fig. 1.1, after the orientation of our two images we have reconstructed the complete geometry. This means that if we have three-dimensional object (terrain) co-ordinates of a point inside of the model area, it is possible to calculate the pixel co-ordinates of this point in the left and the right image using the well-known *collinearity equations* (see below). The programme uses this fact in the way that for each point of our control point file CONTROL.DAT a test is made if this point is represented in both images. As a result, the model area can be calculated (section **Range**).

Collinearity equations (see ALBERTZ & WIGGENHAGEN 2005 for instance)

$$x' = f * \frac{a_{11}(x-x_0)+a_{21}(y-y_0)+a_{31}(z-z_0)}{a_{13}(x-x_0)+a_{23}(y-y_0)+a_{33}(z-z_0)} \qquad y' = f * \frac{a_{12}(x-x_0)+a_{22}(y-y_0)+a_{32}(z-z_0)}{a_{13}(x-x_0)+a_{23}(y-y_0)+a_{33}(z-z_0)}$$

with (x, y, z) object co-ordinates, (x_0, y_0, z_0) projection centre, f the focal length and a_{11}, a_{12}, ... the coefficients of the rotation matrix.

$$(4.3.1)$$

By comparing the distances between different points in object and pixel co-ordinates, the mean pixel size of our aerial images in terrain units [m] and the approximate scale of the original photo are determined. The ratio between flying height and image base (=distance between both projection centres, see Fig. 1.1) is an indicator for the maximum attainable accuracy in z, calculated as a product of the first two parameters (section **Geometry**).

If the orientations of our two images would be exact, then each individual object point within the model area together with both projection centres should form a so-called *epipolar plane* (see Fig. 1.1). In particular, this means that the intersection points of the projection rays [object point → projection centre → film plane] are homologous; there are no y parallaxes. But, as you remember, nothing is exact in real life! And indeed, due to the geometrical situation of the scanner or the CCD sensor of a digital camera, geometrical resolution of the image, errors in the fiducial marks and control point co-ordinates as well as their measurement results and other influences, usually even in a completely oriented model we will have y parallaxes in a range of some pixels. This will be disturbing for viewing, interpretation and automatic processing steps. Therefore, the programme can consider y-parallaxes in the selected range (for instance ± 1 pixel).

Created file: 157158.MOD

4.4   Stereoscopic Viewing

From now on we are able to view the model stereoscopically, measure three-dimensional object co-ordinates and digitise objects like points, lines or areas, sometimes called *feature collection*.

Before going on with practical exercises, let's talk about some basics of stereoscopic viewing. May be that you are more familiar with the term "stereo" in association with music: It gives you a spatial impression like sitting in front of an orchestra. The reason is that you receive the sound with *two* ears, sounds coming from the right primarily with your right ear, sounds coming from the left primarily with your left ear. As a result, both acoustic signals are slightly different and are combined in your brain to obtain a spatial impression.

From this well-known experience we can directly go over to stereo vision. We see the world around us with *two* eyes, and due to the fact that they receive the optical information as a central perspective with a distance of about 6.5 cm (about 2.6″) between the two images, they are slightly different. These again are combined in our brain to obtain a spatial impression.

This is *one* important reason why we are able to estimate distances. But it is not the only one. Perspective—farther objects seem to be smaller than nearer ones—, experience, background knowledge and more helps us to get a spatial view. May be you know this test: Put a glass of beer on a table in front of you, close one eye, an try to touch the border of the glass with a finger… Repeat the test with both eyes opened (before drinking the beer…).

So far about real life. In our case we would like to see *images* of the real world with the same spatial impression like the world itself. The problem is that to reach this goal we must take care that the left image will only be seen by the left eye, the right image only by the right eye. Once again remember stereo listening: The best way to separate both *channels of acoustic information* is to wear headphones. Analogue to this, the best way to separate both *channels of optic information*, the left and the right image, is to use special optics like a mirror stereoscope to view paper images.

For the stereo vision of digital images there exist a couple of possibilities. For example, the images can be viewed with so-called *shutter spectacles*, having a small LCD screen for each eye. Here, a high speed switching between left and right image on the screen and simultaneously the left and right part of the spectacle will be carried out—at each time, only one glass of the spectacle is switched to opaque. Or, each image is projected separately, one horizontally polarised, the other vertically. In all cases you need special hardware. A very interesting method is to use a special colour coding based on the psychological effect that, for instance, red objects lying in the same plane as blue objects seems to be closer to the viewer. Simple prism spectacles can enhance this effect (see Petrie et al. 2001, for instance) (Fig. 4.8).

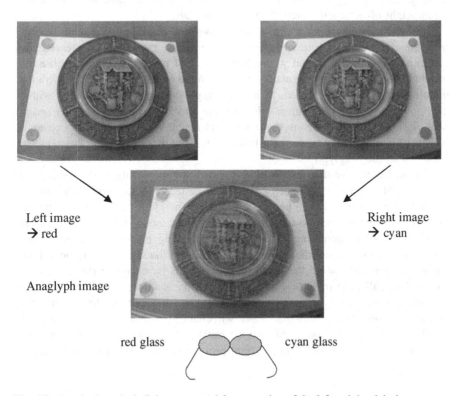

Left image
→ red

Right image
→ cyan

Anaglyph image

red glass   cyan glass

Fig. 4.8 Anaglyph method. Colours are used for separation of the left and the right image

A very simple and more than 160 years old method is often used for viewing stereo printings and is also used here. It is called the *anaglyph* method (developed by Rollmann, 1853). The idea is to print or display both images overlaid, the left in one base colour (usually red), the right in a complementary colour (usually cyan). Wearing a simple spectacle with a red glass left and a cyan glass right, acting as filters, your left eye will only see the left image, your right eye only the right one. The advantage of this method can be seen in the costs, no special hardware is necessary, and in the fact that several persons can use this cheap method simultaneously. The only disadvantage is that in colour images red and cyan objects appears to be grey because these colours are used for separation of the two images, as described.

4.5   Measurement of Object Co-ordinates

Please start **Processing > Stereo measurement**. Until now, we have no terrain model of our area, and consequently the option **or z =** is suggested, the start or initial height here been set by the programme to the mean z value of the height range given in the project definition (ca. 1350 m). Set this value to 1200 m, the mean height of our central area, then click onto **OK**.

The stereo display appears with a similar look and handling like already known from the orientation measurement. Holding the middle mouse button pressed you can move the model in x and y direction. For rapid movement, you may use the rectangle border (showing the actual position) in the overview image, also with the middle mouse button depressed. In fact, moving the mouse means moving within the *object space*: The programme calculates the intersection points (as described in Sect. 4.3) using the collinearity equations and is able to move the images simultaneously. Viewing may be done for both images side by side or using the red-cyan glasses with the anaglyph method.

You will recognise that the positions "under" the measuring marks for the left and the right image are not exactly identical. The reason is that with exception of our control points we have no height information. Therefore we must set the height manually: Press the right mouse button and move the mouse forward (=increasing z) or backward (=decreasing z) until both measuring marks lies "over" corresponding positions, or use the central mouse wheel. In the status line bottom left on the screen you can pursue the result of your actions by changing the x, y and z values. Eureka![1]: The main task of photogrammetry is solved!

Attention, theory: If you are a sharp observer you will recognise that, moving the mouse in z direction, not only the z value will change but also the x and y values by small amounts. The reason is that we are moving within the *epipolar plane* (see Fig. 1.1) defined by the left and the right projection ray, along a vector from our

---

[1]Greek: I have found it!, © Archimedes, 3rd century before Christ.

actual start position x, y, z (terrain) to the middle of both projection centres, and that we have no nadir images (see Sect. 1.7).

Obviously, now we are able to measure (digitise, register, collect) the three-dimensional object co-ordinates of every point within our model, a work which is sometimes called *feature collection*. You can imagine a lot of applications for this tool, let's mention only two of them here:

- Cartography and GIS: Digitising roads, rivers, buildings and similar objects and use these data for example within a cartographic or GIS software package.
- Terrain models: Collection of points and morphologic data like break lines, using these data to interpolate a DTM (see Sect. 4.6.1).

For both possibilities we want to give a little example here. As you can see on the display, the central part of our model shows the beautiful city of Caicedonia, a typical Spanish founded settlement in Colombia (South America) with a chessboard-shaped ground-plan. Let's digitise some of the housing blocks, called *cuadras* (squares), with a surrounding polyline. To do this, choose **Register > Points/lines** from the main menu. In the appearing window set the parameter **Code** to **General lines** and the name of the output file ("Save as") to REGIS.DAT, then click onto **OK**.

Now go near to the first point which should be a corner of a square, using the middle mouse button, set the correct height with the right mouse button pressed down, correct the position and then click onto the left mouse button. Go to the next point, set the height, and so on. Close the polyline with a click onto the **Close** button right hand in the **Register** menu segment, in this way creating a polygon (=closed polyline). A window appears in which you may change the code (don't do it here) and go to the next polyline with **Continue**, or finish the measurement with **Ready**—in the latter case, the programme will inform you about the number of digitised points.

You will feel that this kind of digitising is a bit complicated and time-consuming: In more or less any position setting x and y, then correcting the height, fine correction of x and y, in some cases final correction of z... You will agree: There must be a way to make the work more comfortable. Indeed, there are several ways, and in our next example we will see the first of them:

Choose **Register > Grid** from the main menu. With this option it is possible to digitise a regular grid of points semi-automatically in the way that the x-y-positions are set by the programme, and the user has only to correct the height. In the window you see the proposed area (range of x and y). Please key in the following parameters:

```
X from 1138000 to 1140000
Y from 969900 to 971500
Grid width 250 m
```

Set the output file name again to REGIS.DAT, then click onto **OK**. May be a warning message "File already exists" appears—in this case click onto **Overwrite**.

Now the programme sets the images to the first position, given by the minimum x and y values of your grid's border. Set the height using the right mouse key or the central mouse wheel in the same way like in the last example, then click onto the left mouse button. The programme goes automatically to the next position, you just have to set the height and click onto the left mouse key and so on until the last position is reached (=the maximum values of x and y). As you will see it may happen that a position is a bit outside of the model area or that a point cannot be measured, for instance because it is covered by a cloud. Then go to the next position by clicking onto the Skip button in the Digitise menu segment or simply use the F3 key until the next position is reached where a measurement is possible. Continue until the last point is measured, or click onto Ready to finish this example.

Remark: In any of the measurement windows, you can use Info > Mouse buttons to see the options associated with the mouse buttons and function keys.

Obviously this is a way to make the digitising easier if the goal is to collect regularly-spaced data. Let's keep it in mind to look for more comfort also if we want to digitise individual objects like lines...

More or less all we have done until now is also possible with analytic instruments—of course, they are much more expensive and complicated to handle. But, why not use some of the powerful tools of digital image processing to get even more automatic and comfortable? In the next chapter you will learn something about the possibilities of image correlation (matching). With this step we will enter the field of methods which are typical for digital photogrammetry and not available in (traditional) analytical work.

Created file: REGIS.DAT

4.6 Creation of DTMs via Image Matching

4.6.1 Some Theory

Please remember what we have done just before: Automatic pre-positioning in x and y, then manually setting the height. We considered the height to be OK just in the moment when the measuring marks were lying exactly "over" corresponding (*homologous*) positions in the left and the right image. If we can find an algorithm telling the computer what we mean saying "corresponding positions", the programme should be able to do all the work automatically over the whole model area, forming a digital terrain model (DTM) (Fig. 4.9).

In general, a DTM can be seen as a digital representation of the terrain, given by a more or less large amount of three-dimensional point co-ordinates (x, y, z). There exist various methods to get these data, one of them we will see in the next chapter. From the input (primary) data, may they be regularly distributed or not, then an area-covering data set (secondary data) is created by interpolation (for instance, see Linder 1994).

Fig. 4.9 Homolgue points (*white arrows*). Take care of moving objects like ships! Area: River Rhein near Düsseldorf

Before going on, this is the moment to remember what kind of elevations we or the programme are able to measure: For each given position (x, y) the uppermost z value and only this! Depending on the land use, this may be directly on the terrain but also on top of a house or a tree. For instance, image parts showing a dense forest will lead to a surface on top of the trees. Therefore, we know two different definitions of elevation models (see Fig. 4.10).

- Digital terrain model (DTM) or sometimes digital elevation model (DEM): Contains z values situated on top of the real terrain (earth). Such a model can be used to derive contour lines.

Fig. 4.10 Situation in the terrain and kinds of digital elevation models

What we have: The real situation...

... what we can get by image matching: Surface model, needed for the production of ortho images

... and what we would like to have sometimes: Terrain model, needed for the derivation of contour lines

- Digital situation (or surface) model (DSM): Contains z values at the top of objects situated on the terrain. Such a model is needed when ortho images should be created (see Sect. 4.7).

During the past, a lot of efforts have been made developing methods for this. We will use one of them, well-known as *area based matching* (ABM) which in general leads to good results if we have good approximations. To give you an idea about this approach, please take a look at Fig. 4.11.

The programme compares the neighbourhood of a point in the left image (sometimes called *reference matrix*) with the neighbourhood of the approximate corresponding position in the right image (sometimes called *search matrix*), moving the right position and matrix for a given number of pixels left-right and up-down. In any position a value is calculated giving a measure of correspondence. For this the *correlation coefficient* has been proven to be useful.

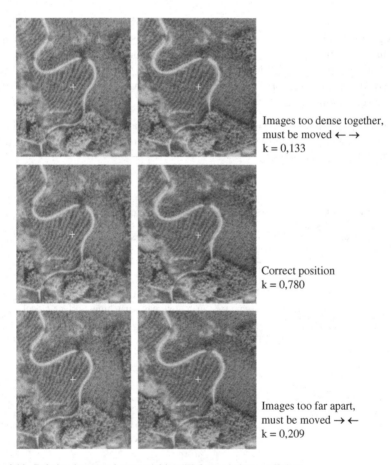

Images too dense together, must be moved ← → $k = 0,133$

Correct position $k = 0,780$

Images too far apart, must be moved → ← $k = 0,209$

Fig. 4.11 Relation between image positions and correlation coefficient

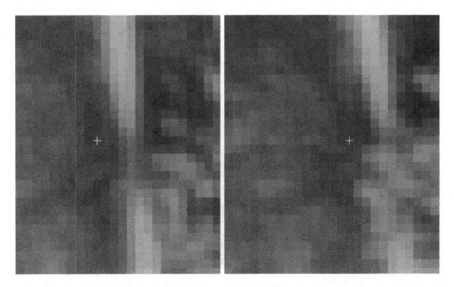

Fig. 4.12 Parts of the left and the right image, strongly zoomed. The grey values are similar but not identical. Therefore, the correlation coefficient will not be equal but will be near to 1

As you may know, the absolute value of the correlation coefficient ranges between 0 meaning that both pixel matrices are completely different, and 1 meaning that they are identical. So, the programme will recognise the correct corresponding position by the maximum of all coefficients. For more details about this topic see Hannah 1988 or Heipke 1996, for instance (Fig. 4.12).

The movement of the search matrix, displacing pixels left-right and displacing pixels up-down as mentioned just before, is the standard method in most matching programmes because they are working in the *image space*. The software we use here works in the *object space* as we already have seen before, a technique called *vertical line locus* if we have more or less nadir images (see also Fig. 1.1). As a consequence, we have no classical reference or search matrix. The programme simply moves within a given interval along the vector from our actual start position x, y, z (terrain) to the middle of both projection centres (see Sect. 4.5), as a consequence the intersection points between the projection rays and the image planes are moving along epipolar lines, and neighbourhoods of the intersection points serve as reference/search matrices.

So far theory was covered. In practise, there may occur a lot of problems due to the fact that the programme has to compare parts of two *different* images showing the *same* object from *different* positions. Besides different brightness and contrast conditions we remember that the relief leads to radial-symmetrical displacements of the objects (see Fig. 4.13). As a result, neighbourhoods of corresponding points in neighbouring images are normally not completely identical or, in other words, the correlation coefficient will never reach the value 1. Nevertheless, in areas with good contrast and flat terrain values of more than 0.9 may occur.

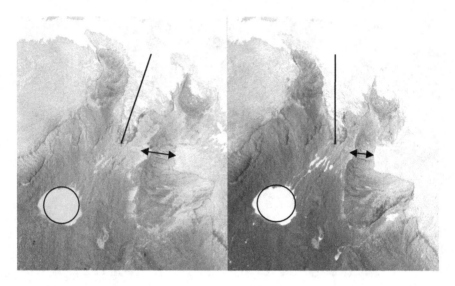

Fig. 4.13 Displacements caused by the relief, grey value differences from reflections. Area: Nevado de Santa Isabel, Colombia

On the other hand, in areas with low contrast and/or strong relief like high mountainous regions it can happen that the correlation coefficient will not be higher than 0.3 even in exactly corresponding positions. Concerning the influence of the relief you can imagine that the size and the form of the neighbourhoods to be compared also have an effect on the results: For instance, a smaller matrix will show a smaller influence of the relief. But we cannot establish the rule that smaller matrices are better as we can see looking at a problem known as "repetitive structures" or "self-similarity":

Imagine a field with crops, let's say potatoes, usually situated in parallel rows. Depending on the aerial photo's scale and the scan resolution, a matrix of 21 by 21 pixels may contain two or three rows, and moving the search matrix over the field you will find a lot of similar samples all giving a relatively high correlation coefficient. If we increase the size of the matrices to 51 by 51 pixels we will have a better chance to get the correct result because the programme may find enough small differences (see also Fig. 4.11).

As a general rule we can say that large matrices are more stable but less accurate, small matrices are less stable but, *if* the corresponding position is found, more accurate (for instance, see Hannah 1989, Jordan/Eggert/Kneissl 1972 or Mustaffar & Mitchell 2001).

4.6.2 Practical Tests

Enough of theory for the moment—let's begin with an example. Please start Processing > Stereo correlation. In the next window control and/or set the following parameters:

Px ± 3 pxl, Correlation coefficient r > 0.7, window 11 pxl, 3 iterations. Image pyramids activated, Interpolation de-activated. Files: Object co-ordinates CONTROL.DAT, Output GITT_TST. IMA, maintain all other parameters as given by default.

Be a bit patient, because this step will need some seconds. After the approximate DTM is created, you can observe on the screen the results of the improvement as a graphical preview, more and more filled in every iteration.

Remark: There are gaps in the DTM because we deactivated the option Interpolation (Fig. 4.14).

Now choose Palette > Colour 1 to get a better impression of the terrain structures—as you can see, even the streets and buildings of Caicedonia were found by the programme. Most parts of the model area are covered with pixels, nevertheless some gaps are remaining. Please close the display window, for example using the Esc key. The size of the gaps or, in other words, the amount of DTM pixels on which the correlation failed, may be reduced by several methods:

Fig. 4.14 DTM derived from image matching

- Increase the px range: But this only makes sense if the terrain is mountainous.
- Decrease the correlation coefficient threshold: We can do this in images with good contrast and low relief influences like here, but be careful in other cases.
- Decrease the size of the correlation window to reduce relief influence (see also Sect. 4.6.1).
- Increase the number of iterations.

Which parameter(s) we may change depends on many aspects, for instance the accuracy we want to get—remember our discussion in the last chapter concerning the correlation window size, for instance. Because of the fact that we have moderate terrain and images with good contrast, let's try the following in our case:

Start **Processing > Stereo correlation** again and change the following parameters with respect to our first attempt: **Correlation coefficient 0.7, Correlation window 7 pxl**. Maintain all other parameters, set the name of the output file again to GITT_TST.IMA and start the matching process. After the calculation has finished, you will see that in more pixels (DTM positions) than before the correlation process was successful.

For a first evaluation of the results, we use the following method: You already know the stereo measurement option—so start **Processing > Stereo measurement**. In contrary to Sect. 4.5, now we have a DTM (GITT_TST.IMA) which is proposed to be used in the first window. Click onto **OK**.

When the stereo display appears, move a bit inside of the model with the middle mouse button depressed. You can see that in nearly all positions or, specifically, in positions with known height, the corresponding image parts fit together perfectly. In the status line lower left the z value is changed dynamically during the movement according to the DTM position.

Activate **Overlay > DTM points** from the main menu with the result that all DTM positions are superimposed red-coloured in the left and the right image.

Let's keep in mind that the amount and quality of the correlated points depends on the quality of the images and the image orientations, the correlation coefficient limits, the window size and the number of iterations. We will discuss some more aspects of quality in Sect. 4.6.4.

In standard cases, the option **Interpolation** is active and the gaps are filled. But, for instance, if you have images with larger areas of very low contrast and as a result larger areas with no correlated points but only filled via interpolation, the quality of the final DTM in this areas may not be good. To handle this problem you have the possibility of measuring additional points manually and include them into the DTM before the interpolation of gaps is carried out, as we will see in the next chapter.

Created file: GITT_TST.IMA

4.6.3 Additional Manual Measurements

Stay within the stereo measurement interface, GITT_TST.IMA loaded as DTM. Start Register > Points/lines, use Code = General points and set the name of the output file ("Save as") to ADDI_PNT.DAT, then go on with OK.

Now measure points in areas were the correlation failed (see Sect. 4.5). In particular consider peaks on mountains, local depressions and similar forms deviating from plane terrain. If possible, measure also points in areas with low texture. And finally, you may also register linear structures in the images with the code "break lines" (see Appendix 1).

Remark: The programme will use height (z) information for the pre-positioning if available from the stereo correlation and maintain the last actual height in gaps. As an alternative you may activate the Z—option (z = constant) with the result that the actual value will be maintained independently from existing heights in the DTM.

After the last point is measured, the programme stops giving you information about the number of measured points.

Now you can join the DTM with gaps (raster image) and the file just created (vector data) to obtain a final DTM: Select Processing > DTM interpolation and set the input files: DTM raster image GITT_TST.IMA, vector data ADDI_PNT. DAT, activate actual model and a 5 × 5 mean filter, then click onto OK. The result is a DTM without gaps.

This is the moment to remember Sect. 4.5, digitising individual objects like lines in the stereo model: We found it not to be efficient that in any point we first had to set the position in x and y, then set the height, correct the position etc. If you like, you may repeat Sect. 4.5, using our DTM instead of a (constant) start height in the first window. Now, if you move around in the model or if you digitise points or lines, the height is most always set to the correct value, and it is much easier to collect data. Another possibility will be presented in Sect. 4.7.5.

Created files: ADDI_PNT.DAT, GITT.IMA

4.6.4 Quality Control

It would be good to get an idea about the quality of the DTM derived by matching. In general, we must divide between interior and exterior accuracy:

The *interior accuracy* can easily be controlled using the option Processing > Stereo measurement which you already know, loading the model and the DTM, then moving around in the model area and observe the positions of the measuring marks in the left and right image. If they are at all times in homologous positions, the interior accuracy is good—a manual measurement will not lead to better results. Or, use the Overlay > DTM points option and control the points in both images.

Depending on the quality of the image orientations, nevertheless it is possible that the corresponding terrain (object) co-ordinates are not very accurate. To check this—let's call it *exterior accuracy*—we need an independent reference.

For this you can use any kind of 3D-information about the terrain which are *not* generated from our aerial images. These may be point data for instance from a GPS campaign in the field, or contours from a digital map, available as simple vector files (x, y, z per point). Within the stereo display load such files using the Overlay > Vector data option and control the positions like described before.

4.7 Ortho Images

It is a simple work to geocode or rectify a digital (scanned) topographic map. Just search a few control points (x, y), measure their positions in the map, and use a simple plane affine transformation (see Sect. 4.2.2; you can do this for instance with LISA BASIC). The reason that such a simple 2D approach leads to good results is the fact that the map was created with a so-called *stereographic projection* where all projection rays are parallel and rectangular (orthogonal) to the projection plane.

But if we want to rectify an aerial image we have to deal with some problems, most of them resulting from the (natural or artificial) relief and the central perspective projection, leading to radial-symmetric displacements (see Sect. 1.6). These are pre-requisites for stereoscopic viewing and 3D measurement as we saw before but make rectification more complicated. The solution is called *ortho image*, a representation in the same projection like a topographic map, and again we will start with some basic theory.

4.7.1 Some Theory

Please take a look at Fig. 4.15. If we have one or more completely oriented image (s) and information about the terrain surface—like our DTM created before—, then the only thing we have to do is to send a ray from each image pixel through the projection centre down to earth. The intersection of the ray and the terrain surface gives us the correct position of the start pixel in our output image. This process, carried out pixel by pixel, sometimes is called *differential rectification*.

The theory is easy, but even here problems may occur. For example, very steep slopes, may they be natural or artificial (walls of houses etc.), will lead to hidden areas in the image. Obviously this effect increases with stronger relief, greater lens angle and greater distance from the image centre—the effect is not only a hidden area in the image but also a gap in the ortho image. On the other hand, objects situated in or close to the terrain nadir point will have no or nearly no displacement in the image.

Fig. 4.15 Central projection (images) and parallel projection (map, ortho image)

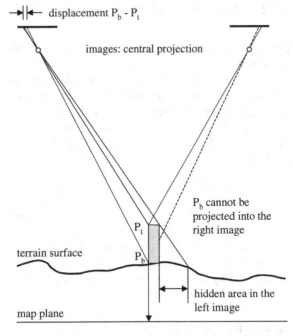

displacement P_b - P_t

images: central projection

P_b cannot be projected into the right image

P_t

terrain surface

P_b

hidden area in the left image

map plane

map or ortho image: parallel (stereographic) projection

Representation of the building...

...in the left image   ... in the right image   ...in the map

There exist several methods to handle the situation. First, we will change the direction of the projection rays: [ortho image → terrain surface → projection centre → image], sometimes called the *indirect resampling method*. By this process we will get a grey value for all ortho image pixels, no gaps will occur. Second, in the case that we have more than one single image (for example a stereo model), we will follow the rule "nearest nadir" which means that we will use *that* image in which the corresponding point of our actual object position is situated as near to the image centre as possible (see Miller and Walker 1995, for instance).

Please note that the geometric accuracy of an ortho image is highly dependent on the accuracy of the DTM: Let's take an object point with its representation near to the image border (=far away from the nadir point), the image taken with a wide angle camera, then a height error of dz will lead to a position error of more or less the same size!

Therefore you can easily imagine that an optimal rectification must be done with a digital situation model (DSM), and that's just what we got via stereo correlation.

4.7.2   Resampling Methods

Using the rays [ortho image → terrain surface → projection centre → image] like mentioned before, in any case we will find an aerial image pixel for each DSM pixel. But, as you may imagine, normally the aerial image pixel matrix will not be parallel to the ortho image pixel matrix and the pixel sizes of both images will differ. This leads to the question how to handle the resampling process in particular.

Figure 4.16 illustrates what we mean. There exist several methods to determine the grey value for the new (ortho) image, and each of them has typical advantages and disadvantages:

- Nearest neighbour: This method is the fastest one. If the geometric resolution of the aerial images is significantly higher than that of the ortho image, select this option. In our example (Fig. 4.16), the ortho image pixel will get the grey value of pixel 3 from the original image.
- Bilinear: This method may be used if the geometric resolution does not differ very much. The bilinear approach calculates the new grey value from 4 pixels surrounding the intersection point, weighted by the distance from the

Fig. 4.16 The resampling problem: Find the grey values for the pixels in the new image

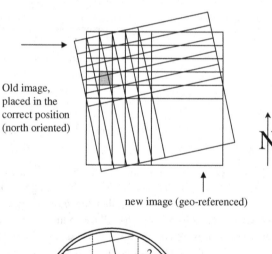

Old image, placed in the correct position (north oriented)

new image (geo-referenced)

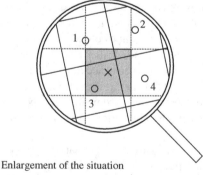

Enlargement of the situation

intersection point (marked with x) to the pixel centres (marked with o). This leads to a "smoother" result, due to the fact that the resampling method has a mean filter effect! Like for every mean filter, the resulting grey values are not identical with the original ones. This must be taken into account if the result should be classified afterwards.

- Cubic convolution: Similar to bilinear, but 16 surrounding pixels are used, leading to a stronger smoothing (and more time for calculation). This method is not offered in our software.

4.7.3 Practical Tests

OK, enough of words, let's start with LISA FOTO again and go to Processing > Ortho image. The parameter Source gives us three alternatives: Single image (the exception), Actual model (default and our case) or All images which we will use later on (Sect. 5.3.3). What about the other parameters?

Colour value adjustment: If the same object area is represented in different images it often happens that there are differences in brightness. Sometimes, especially in older photos, you will recognise a brightness decrease near the borders (*vignetting*). So, if we will calculate a mosaic of ortho images we may get sharp brightness changes along the seams which are the "nearest nadir" borders of

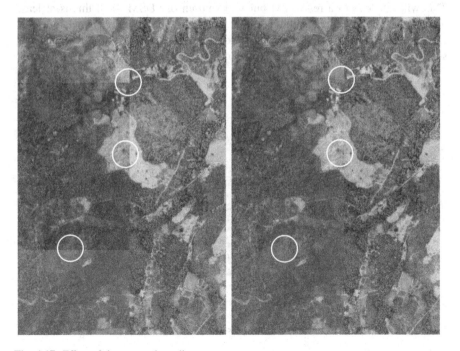

Fig. 4.17 Effect of the grey value adjustment

neighbouring images mentioned before. Using the option described here, the programme will calculate a correction function for every image with which brightness differences will be minimised (Fig. 4.17).

Resampling: Bilinear.

Files: For the DTM select GITT.IMA. As an alternative which we will not use here it is possible to create the ortho image using a horizontal plane (z value constant) instead of a DTM.

Click onto OK. After the programme has finished, the result will be displayed (see also Fig. 4.19).

Created file: ORTHO.IMA

4.7.4   *Creation and Overlay of Contours*

As you may know it is possible to derive contour lines directly from a DTM, and that's what we want to do now. But stop, a short brainstorming will be helpful before: What we have until now from the stereo correlation is not really a DTM but a DSM (see Sect. 4.6.1) containing more or less the real surface structure with a lot of peaks like houses, trees and others (Fig. 4.18).

Imagine we will use this to calculate contours, then we will get really a lot of lines running around this peaks, and this is surely not what we want. Contours are always created from DTMs! Therefore we need to filter the DSM with a mean filter. This will not lead to a real DTM but will smooth our DSM—and this is at least more than nothing. In fact, during the past and for instance in connection with DTMs derived from laser scanning, efforts have been made to develop more sophisticated filter strategies (see Jacobsen 2001 or Lohmann 2001, for instance).

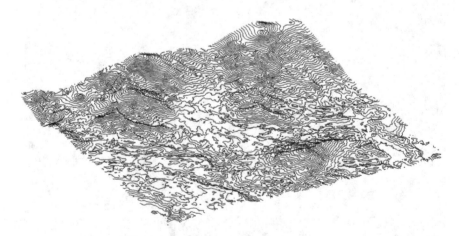

Fig. 4.18 Example of contour lines in 3D representation

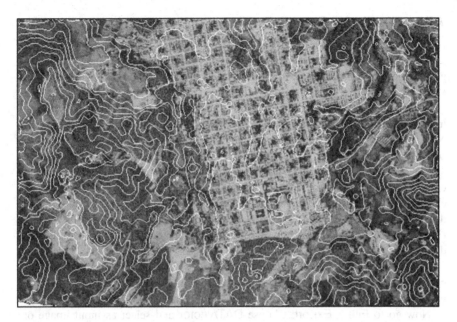

Fig. 4.19 Ortho image, 10-m contours overlaid

For this and the contour creation we will use the LISA BASIC module. Please start this programme, then go to **Terrain models > Filtering**. We will use all options of this tool and the following parameters:

Fill local minima with a window size of 30 pixels, **Remove peaks** with a tile size of 10 pixels and a threshold of $z = 2$ m and **Filter** with the method **Gauss** and a window size of 5×5 pixels. Input file is GITT.IMA, output file as suggested GITT_FLT.IMA. Click onto **OK**.

Now start **Terrain models > Contours vector**. Set the parameters **Equidistance** to 10 meters and **Tolerance** to 1 m and use as name for the output file CONTOUR.DAT. After **OK** the programme starts creating the contours in a vector representation, and when this is done, a data reduction procedure (*tunnelling*) will be carried out. The result will be displayed in the graphics editor of LISA BASIC. Especially when enlarging the graphics you may find several arte-facts—as mentioned above, we have no real DTM, and these artefacts result from peaks not completely eliminated by the filtering process.

Stay within the LISA BASIC programme and display the ortho image, then choose **Overlay > Vector graphics**. Input file is CONTOUR.DAT, overlay colour for instance 255 (white) or 1 (black), as you like. After **OK**, the contours are shown overlaid to the ortho image (see Fig. 4.19).

Created files: GITT_FLT.IMA, CONTOUR.DAT

4.7.5   Simple 3D Data Collection

Let's remember the topic of data collection (digitising). There is a quite easy way to get three-dimensional data if we have an ortho image and a DTM, and we want to explain this here:

Use again the LISA BASIC module and display the ortho image ORTHO.IMA in the standard image display (Display > Raster image or with the popup menu). Now go to Measure > Register, set the code to General lines, Z value keep as zero, set the name of the output file to REGIS_2D.DAT, then OK.

Now you can digitise points and lines very simply in a mono image, just clicking onto the desired positions with the left mouse button and using the Close or the checkmark button right-hand in the window to finish a line—see the programme description of LISA BASIC for more details. When you are ready, click onto the Ready button in the window always appearing at the end of a line, then close the display.

Until now, we have only collected the x and y co-ordinates of our objects. We can control the results using Display > Text, finding that all z values are set to 0, or using Display > Vector graphics to see a graphical representation.

Now go to File > Export. Choose DAT/vector and select as input image our DTM, GITT.IMA, then OK. In the next window choose Single points, point file is REGIS_2D.DAT, and set the name of the output file to REGIS_3D.DAT, again OK. What's going on?

The programme reads the x and y values from the file REGIS_2D.DAT, finds the corresponding position within our DTM and adds the z value to it. Therefore, the output file REGIS_3D.DAT has similar contents like REGIS_2D.DAT, in particular the x and y values are identical, but the z values now give us the DTM heights.

Let's summarise what we have seen during data collection:

- If you have to digitise 3D data from a stereo model, you can do this directly within the stereo measurement option.
- If the amount of data to be digitised is large, it is much more comfortable first to derive the DTM, then loading it into the stereo measurement module to get an automatic setting of the z value in any position.
- It is also possible to create an ortho image from our (original) image(s) and the DTM, then measure data very simply in 2D, after that adding the heights like described above.

And what's the best way? The last method should only be used if the DTM is very precise and/or the heights must not be of high accuracy. Remember that the DTM creation by matching may fail in some areas, for instance due to low contrast—therefore, if you need really good elevation values in all positions, the second method is the better one because you see if points are homologous, and if they are not, you can correct this directly during digitising.

Besides, photogrammetric experiences show that a maximum accuracy of digitising especially concerning the height will be reached in a "real" stereo measurement, for instance using the anaglyph method instead of the side-by-side representation of our images.

Created files: REGIS_2D.DAT, REGIS_3D.DAT

Chapter 5
Aerial Triangulation

5.1 Aerial triangulation measurement (ATM)

If LISA FOTO is still running, go to File > Select project and choose the project TUTOR_2.PRJ or, if you have to start LISA FOTO anew, select this project. If necessary, go to File > Edit project and change the path of your working directory (see Sect. 4.1).

Within this tutorial we will work again with a ground resolution of 5 m to save time and disk space. Of course, if you have time enough, a strong computer and sufficient storage capacity (RAM and hard disk), you can go to a higher resolution of let's say 2.5 m using the option File > Edit project, changing the pixel size to this value.

5.1.1 Common Principles

Remember the exterior orientation in our first example: For both images forming the model we used some ground control points (GCPs) to establish the orientation via a resection in space. To do this we needed at least (!) 3 well distributed points forming a triangle.

Now imagine the case that we have much more than two images, let's say a block formed of 3 strips each containing 7 images as we will use in this example, and we have no signalised points but only a topographic map, scale 1:50,000. Greater parts of our area are covered with forest, so we can only find a few points which we can exactly identify. It may happen that for some images we are not even able to find the minimum of 3 points.

This may serve as a first motivation for that what we want to do now: The idea is to measure points in the images from which we do not know their object co-ordinates but which will be used to connect the images together. These are called *connection points* or *tie points*. In addition, we will measure GCPs wherever we will find some. Then, we will start an adjustment process to transform all measured

© Springer-Verlag Berlin Heidelberg 2016
W. Linder, *Digital Photogrammetry*, DOI 10.1007/978-3-662-50463-5_5

Fig. 5.1 Proposed positions of control points in the block. From Jacobsen (2007)

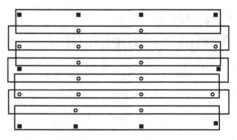

○   vertical control point (only z)

■   x, y, z - control point

points (observations) to the control points. In this way we will only need a minimum of 3 GCPs for the whole block—it is not necessary to have GCPs in each image. On the other hand, a standard rule is to have one GCP in every 3rd model at least near the borders of the block, and if necessary additional height control points inside of the block (see Fig. 5.1).

May be you can get a better impression of this with the help of Fig. 5.2. All images of a block are connected together using corresponding points, "gluing" them to a mosaic which is then transformed to the GCPs. Of course this two-dimensional scheme is *not* exactly what an aerial triangulation will do—nevertheless, it is very useful to understand the rules we must fulfil in our work. The aerial triangulation, today usually carried out in form of a bundle block adjustment (see Sect. 5.2), can be seen as a method to solve an equation system, containing all measured image co-ordinates as well as the GCP terrain co-ordinates.

Remark: After the interior orientation of all images (next step), we will take a look into the principles of manual *aerial triangulation measurement* (ATM) and carry out an example. This is a good possibility to understand the way how ATM works, and is necessary for the measurement of ground control points. Nowadays usually automatic approaches are used, and we will do this in Sect. 5.1.8 as well. But even for understanding the problems or errors occurring in the automatic processing it is valuable to know the basics of manual ATM.

5.1.2   Interior Orientation

Before starting the measurement, we need the interior orientation of all images used in the block. You already know this from our example before—if you want to make it now, please refer to Sect. 4.2.2 and take into account that the first two strips, images No. 134 … 140 and 155 … 161, are taken in "normal" order whether the last strip, images No. 170 … 164, were taken when "flying back". This was considered when importing the image files from the scanner, and you will see it during the sequence of pre-positioning the fiducial marks. The images from our first example are parts of this block, therefore the camera definition is the same as before.

"Glue" the images together to a block...

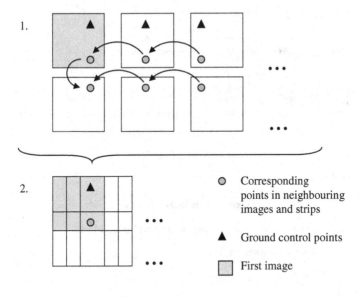

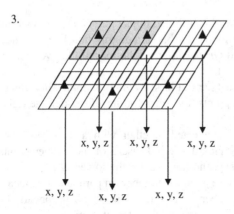

...then transform the block to the ground control points

Fig. 5.2 Scheme of a block adjustment

May be you will find it easier to use the already prepared files, 134.INN ... 170. INN from the Springer server (...\tutorial_2\output).

5.1.3 Manual Measurement

Let's start this chapter with some general rules (see Fig. 5.3):

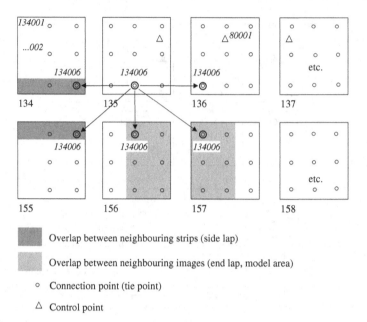

Overlap between neighbouring strips (side lap)

Overlap between neighbouring images (end lap, model area)

○ Connection point (tie point)

△ Control point

Fig. 5.3 Principles of point transfer within a block

- In any *model*, at least 6 well-distributed object points must be measured. It is an old and good tradition to do this in a distribution like a 6 on a dice, the points are then called *Gruber points* in honour of Otto von Gruber, an Austrian photogrammetrist.
- *Neighbouring models* must have at least 2 common points. In standard, we will use 3 of the Gruber points (the left 3 for the left model, the right 3 for the right model).
- *Neighbouring strips* are connected together with at least one common point per model. As a standard, we will use 2 of the Gruber points (the upper 2 for the upper strip, the lower 2 for the lower strip).
- Each object point must have a unique number. In particular this means that a point has the same number in any image in which it appears. On the other hand, different object points have different numbers.

The first step in practice is the preparation of the images. You should have (or make) paper copies of all images. Put them on a table in the order in which they form the block. Now look for the 6 Gruber points within the first model: This should be done in accordance with the advice given for natural ground control points in Sect. 4. 2.4. Take into account that the 3 points at right are also represented in the next model to the right, the 3 points at left are also represented in the neighbouring model on the left. Also take into account that the 2 points on the bottom of the model are represented in the neighbouring strip below of the actual one, the 2 points on the top of the model in the neighbouring strip above the actual strip.

Mark each point in the paper copies for instance with a small coloured circle, and give it a number using a logical scheme: This may be the left image number multiplied with 1000 plus an additional incremental digit. Example: The image number is 134, then you may label the points with 134001, 134002, 134003 and so on. Remember that the stated number of points in the rules are *minimum* values—whenever you find it useful, take more to receive a good connection of images, models and strips.

In the same way mark all existing GCPs, for instance with a small coloured triangle, and give them also a unique number which is different from all others. For example, use the numbers 80001, 80002, 80003 etc.

When everything is prepared, you may begin with the measurement. Start **ATM > Manual measurement**, key in (or control) the image numbers of our first model in the first (uppermost) strip: Left 134, right 135. Set the approximate **Endlap** (= longitudinal overlap between both images) to 65 %, activate the option Create point sketches, set the name of the output file to IMA_COORD.DAT and keep all other file names empty. After OK, a stereo display appears similar to that you already know from the stereo measurement. In the same way you can move both images simultaneously with the middle mouse button pressed down. Using the right mouse button instead you can move only the right image while the left is kept in its actual position (this option is sometimes called "fixed photo" in analytic instruments). In this manner it is possible to put both images together with corresponding points. Please do this in the actual position. It might be helpful to zoom out before by clicking once or twice onto the minus magnifying glass button, then bring both images together, after that set the zoom to standard by clicking onto the central zoom button.

Now start **Measure > Gruber points**. The programme sets both images to the first Gruber position, top left. Hold the middle mouse button pressed, look for a useful position near to the one you are now (for example, a corner of a building), set the desired position in the left image, then hold the right mouse button pressed and do the same for the right image. If the corresponding positions are reached both in the left and the right image, click onto the left mouse button to digitise the point. The programme will give the number 134001 automatically and, after moving the mouse a little, go to the next Gruber position, middle left. Continue the described steps until the last Gruber point (bottom right) is digitised, then click onto the Ready button on the right side of the window and close the display, for instance with **Esc**.

Again start ATM > Manual measurement. Use the $\boxed{>}$ button to switch to the next model, 135/136. The output file name keeps IMA_COORD.DAT as suggested, and again the option Create point sketches should be activated. Attention: After OK, the warning message "File already exists: IMA_COORD.DAT" appears —use the Append button as suggested, *not* the Overwrite one!

When the stereo display is ready you will see the positions of 3 Gruber points coming from the model before marked with small green squares in the overview image. If you move to these positions in the main (stereo) window you can also find them overlaid and labelled in the left image—in analytic photogrammetry this is called *automatic point transfer*—, and of course we will measure these points in our actual model. Besides, 3 new Gruber points in the right part of our model must be measured. It is your decision what you would like to do first.

OK, let's begin with the "old" points marked here. Start **Measure > From model before**, and the programme sets the actual position to point 134004 (top left). The position within the left image of course cannot be changed, so just correct the position in the right image with right mouse button pressed down. Something is new: In the small window bottom left on the screen you see a neighbourhood of the point which may help you also to find the correct position in the right image. This was created due to the option Create point sketches we activated. These "sketches", already known from Sect. 4.2.5, are stored in files named in the form <point number>.QLK, for example 134001.QLK ("quicklook").

After the 3 existing points are measured, start **Measure > Gruber points**. Now we already have points in the left part of the image, therefore use the **F3** key (skip) three times until the Gruber position top right is reached. From this position on continue in the same way like in the last model: Set the point in the left image with the middle mouse button depressed, then set the point in the right image with the right mouse button depressed, after that click onto the left mouse button, and so on.

Finish the measurement with a click onto the Ready button, then leave the window. For control purposes, display the output file IMA_COORD.DAT using the button on the right side of the main window. The results should be more or less like the following:

```
134000135     152.910 RMK_1523.CMR
        1   2752.538  1390.028   2736.462   1389.510
        2   1421.543    54.974   1406.990     54.475
        3     82.058  1381.911     64.985   1376.989
        4   1412.492  2715.526   1392.023   2713.014
   134001   1418.000  2563.600    373.000   2562.600
   134002   1536.000  1383.000    433.000   1361.000
   134003   1471.000   291.400    508.000    249.400
   134004   2262.880  2570.600   1335.880   2572.600
   134005   2225.880  1477.000   1255.880   1465.000
   134006   2383.880   277.400   1515.880    249.400
      -99
135000136     152.910 RMK_1523.CMR
        1   2736.462  1389.510   2739.509   1409.530
        2   1406.990    54.475   1414.581     71.081
        3     64.985  1376.989     68.991   1388.488
        4   1392.023  2713.014   1393.083   2726.904
   134004   1335.880  2572.600    490.880   2573.600
   134005   1255.880  1465.000    348.880   1465.000
   134006   1515.880   249.400    694.880    230.400
   135004   2283.600  2507.600   1532.600   2520.600
   135005   2359.600  1411.000   1601.600   1411.000
   135006   2271.600   224.400   1484.600    196.400
      -99
```

Please note the numbers marked with boxes: Here you can see an example of the automatic point transfer we mentioned above.

In this way you can continue with manual image co-ordinate measurement until the last model of our block is reached.

This is the traditional method, identical to that commonly used in analytic photogrammetry. In the past, a lot of efforts have been made to establish automatic methods which are based on image matching techniques similar to our example for automatic DTM extraction (Sect. 4.6). In the next steps we will learn something about this.

Created file: IMA_COORD.DAT.

5.1.4 Automatic Measurement via Image Matching: Introduction

A programme which shall measure image co-ordinates for an aerial triangulation automatically has to deal with several goals and/or problems. These are, in increasing difficulty:

- Find homologous points within a single model
- … in neighbouring models (point transfer)
- … in neighbouring parallel strips
- … in lateral strips or images of different scale and/or date.

The first two goals can be reached more or less easily. But to connect strips, the programme needs some information about their relative position. One possible way is to define the image centre co-ordinates (for instance GPS data from the camera positions), but this may be a problem if you have greater areas covered by forest, repetitive structures, etc.

A different way is to measure some points for connection manually, serving as initial values. For this, we can use a fast and simple way as described in Sect. 5.1.7. And, of course we must measure the GCPs manually and with high accuracy—this is what we will do in the next chapter.

5.1.5 Co-ordinate Input and Measurement of Ground Control Points

As we will see, a huge portion of aerial triangulation measurement can be done automatically. Nevertheless, a few steps remain to be done manually:

- Input of control point object co-ordinates

- Measurement of the ground control points
- Strip definition (Sect. 5.1.6)
- Measurement of strip connections (Sect. 5.1.7)
- … and then: Automatic measurement of image co-ordinates (Sect. 5.1.8).

The input of the control point co-ordinates may be done in the way described in Sect. 4.2.5. Their values are:

```
Point No.           X               Y               Z

    80001    1136080.500     968916.500      1427.800
    80002    1137755.400     969523.500      1212.200
    80003    1135875.000     971998.000      1089.800
    80004    1137860.000     971648.000      1149.000
    80005    1135318.500     974301.400      1056.200
    80006    1137369.500     973844.200      1120.400
    80010    1139516.400     969242.000      1327.200
    80011    1139925.700     971286.900      1118.800
    80012    1139862.300     973097.900      1108.700
    80013    1141648.200     969138.500      1133.800
    80014    1141901.100     973031.800      1080.900
```

Store the data in a file with the default name CONTROL.DAT, or use the prepared file from the Springer server (…\tutorial_2\output).

We will continue with the measurement of the control points. For each model which includes one or more GCPs (and, in our case, this is true for all models), start **ATM > Manual measurement,** key in the image numbers and the name of the output file, here suggested to be CP_ICOOR.DAT. Click onto the **Append** button if the message "File already exists" appears. After the display is loaded, go to **Measure > Individual** and measure the control points in the way described in Sect. 5.1.3. Figure 5.4a, b inform you about the co-ordinates just stored and the precise position of each point. Use the figures in the Appendix to find the approximate positions of the GCPs.

Created file: CP_ICOOR.DAT.

5.1.6   Strip Definition

The next step is the definition of the strips in our block, which means giving the first and the last image of each strip. Start **Pre programmes > Strip definition.** In the

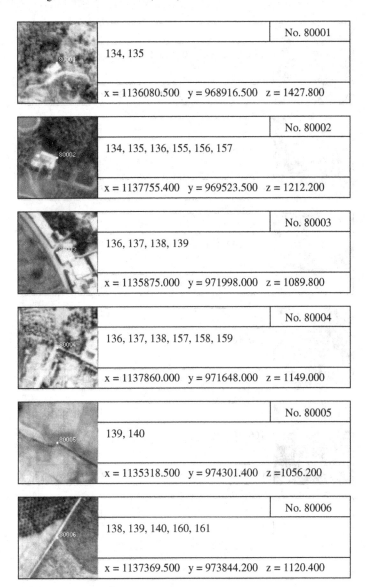

Fig. 5.4 Position and terrain co-ordinates of the control points

appearing, until now empty window click onto the Add button to enter the needed data. In our case, the first and last images are:

```
134         140
155         161
170         164
```

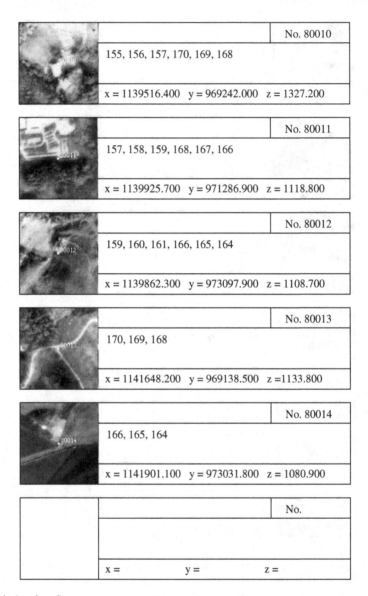

	No. 80010
155, 156, 157, 170, 169, 168	
x = 1139516.400 y = 969242.000 z = 1327.200	

	No. 80011
157, 158, 159, 168, 167, 166	
x = 1139925.700 y = 971286.900 z = 1118.800	

	No. 80012
159, 160, 161, 166, 165, 164	
x = 1139862.300 y = 973097.900 z = 1108.700	

	No. 80013
170, 169, 168	
x = 1141648.200 y = 969138.500 z =1133.800	

	No. 80014
166, 165, 164	
x = 1141901.100 y = 973031.800 z = 1080.900	

	No.
x = y = z =	

Fig. 5.4 (continued)

Fig. 5.5 Part of the graphics interface for the measurement of strip connections. Click with the left mouse button for instance onto the position of this point in all of the 6 images in which it is represented, the sequence is without any meaning. Then, after the last position is digitised, click onto the right mouse button to finish this point and to increase the internal point number by 1

After the last strip was defined, click onto the **Ready** button to store the results. Created file: STRIP.DAT.

5.1.7 Measurement of Strip Connections

The fourth and last preparatory step is to create strip overview images and measure tie points within them: Go to **ATM > Calculate strip images**. For each strip within the strip definition, an image is created showing all aerial images side by side in a size of 800 by 800 pixels. Their names are ST_134140.IMA, ST_155161.IMA and ST_170164.IMA (Fig. 5.5).

Now start **ATM > Measure connections**. In the appearing window, load the first strip (ST_134140) into the upper graphics area and the second strip (ST_155161) into the lower one, each time using the drop-down menus on the right side of the window. You can set the brightness for each strip individually, move each strip with the middle mouse button depressed, and go to the first or last image of the strip with a click on one of the arrow buttons.

Some words before about the measurement process: Digitise a point in all images of both strips in which it appears by clicking with the left mouse button, then click onto the right mouse button to finish the measurement for this point, then do the same with the next point and so on. If necessary, move the strips like described. The points then are numbered automatically and stored in the output file, default name TIEPOINT.DAT, after clicking onto the **Ready** button (checkmark).

Concerning the amount and position of the points, you should follow the rules in Sect. 5.1.3. In particular, make sure to measure enough points situated in neighbouring strips to establish a good strip connection. As an advice, try to measure at least two points in any model to any neighbouring strip.

The results may look like the following:

777770001	621.000	152.000	134
777770001	354.000	154.000	135
777770001	103.000	154.000	136
777770002	720.000	144.000	135
777770002	501.000	140.000	136
777770002	260.000	147.000	137
777770003	705.000	73.000	136
777770003	465.000	85.000	137
777770003	237.000	88.000	138
777770004	740.000	84.000	137
777770004	523.000	81.000	138
777770004	293.000	82.000	139
777770004	67.000	84.000	140

. . .

First column = internal point number, second column = x value, third column = y value (each pixel co-ordinates, measured in the 800 by 800 pixel images), fourth column = image number.

Like before: If you have problems within this step or if you are not sure whether the results are good enough, you may use the file TIEPOINT.DAT from the Springer server, directory ...\tutorial_2\output.

Created files: ST_134140.IMA, ST_155161.IMA, ST_170164.IMA, TIE-POINT.DAT.

5.1.8   Automatic Image Co-ordinate Measurement (AATM)

Now everything is prepared and we can start with the automatic measurement of image co-ordinates, a process sometimes called AATM (*Automatic Aerial Tri-angulation Measurement*). This will need some time, therefore it is a good idea to prepare a cup of coffee.

We will do the complete process in four steps:

- First run of AATM
- First run of BLUH (Sects. 5.2.2 and 5.2.3)
- Second run of AATM to improve strip connections (Sect. 5.2.4)
- Second run of BLUH (Sect. 5.2.5).

Start **ATM** > **AATM** and take a look at the parameters and options in the window: Correlation co-efficient 0.7, window 17 by 17 pixels, 3 iterations. Control the file names: Control points/Image co-ordinates CP_ICOOR.DAT, Control points/Object co-ordinates CONTROL.DAT, Connection points TIEPOINT.DAT and the output file AATM.DAT. Now the coffee should be ready...

Let's use the time for some theory. Internally the images are subdivided into 900 squares—the programme will try to find one point in each square. Depending on the endlap (see Sect. 1.9), the maximum number of points can be calculated. For instance, an endlap of 60 % we lead to a maximum of 60 % from 900 = 540 points. Now remember that we already have several points with known positions in both images: The control points and the points of the strip connection (last chapter). In each of the let's say 540 squares in the left image the programme looks for a position with good contrast. Then, starting with a known point A, a trace to point B is followed in the left image, going the same direction in the right image (see Fig. 5.6).

During the work, the programme will inform you about the progress in an info window, showing the model, the amount of correlated points within this model and to the model before. This may look more or less like the following:

```
===== creating pyramid images =====

strip 1
strip 2
strip 3
```

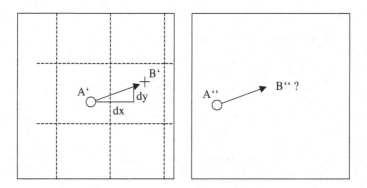

○   Manually measured point (control- or tiepoint)

+   Position with good contrast, found in the left image

⌐   Borders of the interior squares

Fig. 5.6 Automatic search of connection points (tie points) starting with already measured points

From each of the images a second one with reduced resolution is created. It is a well-proved method to start the search of connection points within such smaller images (*image pyramids*).

```
===== searching connection points =====

model                   total    to prev. model

134135                   302          –
135136                   459         130
136137                   500         178
137138                   544         193
138139                   527         224
139140                   536         186

155156                   298          –
156157                   446         118
157158                   504         173
158159                   516         199
159160                   486         188
160161                   476         161

170169                   198          –
169168                   363          71
168167                   486         134
167166                   466         138
166165                   476         134
165164                   461         143

===== 2D pre-check =====

images --> strips
strip  1:           134         140
strip  2:           155         161
strip  3:           170         164

strips --> block
strip  2:      6 corresponding points
strip  3:      5 corresponding points
```

The programme checks whether all images belonging to the same strip can be connected to another, and after that, whether all strips can be connected to a block. If the check was successful, the search for connection points is repeated within the original images (full resolution) to improve the results. Beneath the number of points in each model, the mean correlation coefficient (mcc) is listed:

```
===== original images =====

model                total         mcc

134135               230           0.85
135136               364           0.86
136137               413           0.87
137138               460           0.86
138139               444           0.87
139140               458           0.87

155156               217           0.82
156157               358           0.85
157158               421           0.87
158159               422           0.86
159160               379           0.85
160161               347           0.85

170169               152           0.83
169168               295           0.85
168167               401           0.87
167166               385           0.87
166165               377           0.87
165164               361           0.88
```

Created files: AATM.DAT, DAPHO.DAT (= pixel co-ordinates from the file AATM.DAT transformed to image co-ordinates, necessary for BLUH).

Please note for your own applications that LISA FOTO can only handle parallel strips with images from the same flight (same scale). Until now, lateral strips or images with different scales cannot be processed in the AATM.

5.2 Block Adjustment with BLUH

5.2.1 Introduction

In the next step we will calculate the object (terrain) co-ordinates of all measured image points and also the parameters of the exterior orientation for each image. For this, we will use a so-called *bundle block adjustment* which handles all bundles of rays [object point → image point] together in one adjustment process. In our case, we will take the BLUH software package from the University of Hannover.

Some information cited from the BLUH manuals, copied onto your PC (directory c:\program files (×86)\lisa\text) during the installation, shall give you a first idea about the methods (see also Sect. 10.8.11):

"The bundle block adjustment is the most rigorous and flexible method of block adjustment. The computation with self calibration by additional parameters leads to the most accurate results of any type of block adjustment. Even based on the same photo co-ordinates an independent model block adjustment cannot reach the same quality; this is due to the data reduction by relative orientation, the comparatively inexact handling of systematic image errors and the usual separate computation of the horizontal and the vertical unknowns.

The programme system BLUH is optimised for aerial triangulation but not limited to this. Even close-range photos taken from all directions (with exception of $\omega = 80 \ldots 120$ grads) can be handled. A camera calibration for close-range applications is possible even with special optics like a fisheye lens. Also panoramic photos can be handled in the adjustment.

Special possibilities for the controlled or automatic elimination of a greater number of blunders like it occurs in AATM are included.

The programme system is subdivided into several modules to ensure a flexible handling. For computation of a bundle block adjustment only the modules BLOR, BLAPP, BLIM and BLUH are necessary, they can be handled as one unique set or separately. The other modules can be used for special conditions, for analysis of the data and for other support of the data handling" (Jacobsen 2007).

The principle of bundle block adjustment is based on the collinearity equations, a method to calculate the orientation parameters from ground control points and their positions in the image (see also Sect. 4.3). All point measurements as well as all available control point co-ordinates are handled simultaneously in one single adjustment process which gives the guarantee of high precision results.

5.2.2 Running the Block Adjustment

Please start the programme BLUH_WIN and use the same project like before. The graphics interface is similar to that of LISA, therefore you should not have general problems. The following text will inform you about the single steps and the parameter settings for the BLUH modules BLOR (Pre 1), BLAPP (Pre 2) and BLIM/BLUH (Main; see Fig. 5.7). In the next chapter we will discuss the results.

Select **Block adjustment > Pre 1 (BLOR)**. Check the following parameters:

Photo co-ordinates DAPHO.DAT, Control points CONTROL.DAT, below that line activate the option **Control points**. For the arrangement of the photos in the strips activate the option **STRIP.DAT**. Normally, all these parameters should already be selected when starting this part of BLUH. Now click onto **OK**.

Next window: Maintain all parameters as they are in the section Parameters. Some information about the standard deviations: For the image co-ordinates we can start with a value between 1 and 2 pixels. Remember our scan resolution of 300 dpi which produce pixels of about 84.7 µm size, so key in a value of 100 µm. For the transformation of the strips a value of about 2 pixels is a good start, therefore set it to 150 µm. And finally, the values for the control points depend on their approximate accuracy concerning the terrain co-ordinates and on the accuracy of the

Fig. 5.7 Workflow and interchange files in BLUH. Simplified from Jacobsen (2007)

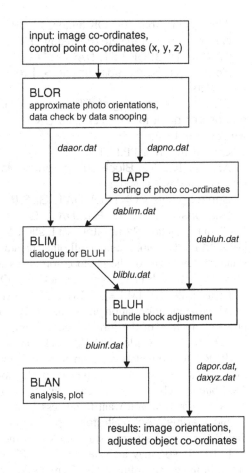

measurements. The image pixel size is about 1.2 m, the terrain co-ordinates have an accuracy of approximately ±1 m, so set the values for x, y and z to 1 m.

Within this first module, BLOR, a check on large errors (blunders) will be carried out using a method called *data snooping*. As a result, an internal file of detected errors is created (DACOR.DAT). Based on this file, we can create an error correction file for an automatic elimination of incorrect measurements in the following modules. The errors are internally classified with asterisks from * = small error to **** = large error. In correspondence to this, the user can set the range of automatic elimination independently for three classes of points (control p., tie p. and others). We suggest to activate the correction option for each of these groups and select the range as following: Control points '****' (delete only large errors), tie points '***' (a medium value to establish a sufficient strip connection) and '*' for all other points (delete even small errors). Because of the fact that we have a really large amount of points in each image we can use '*' for all other points—of course, many of them will be eliminated now, but we have more than enough. This is a typical setting for error correction when points are found via AATM.

After clicking onto the **OK** button, BLOR starts in the background. Finally the results are shown in the editor window. Please close this.

Created file: BLOR.LST, DACOR.DAT.

Now select **Block adjustment > Pre 2 (BLAPP)**. Check the following parameters:

Set the **Limit for listing of differences** to 200 μm and the **Maximum number of photos per point** to 50. Click onto **OK** and close the results shown in the editor window like before.

Created file: BLAPP.LST.

And finally select **Block adjustment > Main (BLUH)**. Again please check or select the following options:

Control points CONTROL.DAT, GPS/IMU positions remains empty, **Photo orientations (output) DAPOR.DAT, Object co-ordinates (output)** DAXYZ. DAT. In the section **Parameters** set **Listing of residuals** to 150 μm, **Warnings in iteration** to 200 mm, maintain all other parameters in this section. Finally, set the standard deviations of the photo co-ordinates to 100 μm and those of the control points to 1 m each in x, y and z. Click onto **OK**.

Created files: BLIM.LST, BLUH.LST, DAPOR.DAT, DAXYZ.DAT.

All protocol information and results of BLUH are stored in simple text files, the protocols in files of the form <module name>.LST, for instance BLOR.LST, the final results in the file DAXYZ.DAT (adjusted object-points co-ordinates) and DAPOR.DAT (orientation parameters of all images in the block). If you like you may open a file using the popup menu (right mouse key) or just by clicking onto the **Text** button right side in the main window to open the text editor.

Two remarks if you want to process your own datasets:

Block adjustment > Strategy opens a tool which will set several parameters in the BLUH modules to realistic values. See the program description for details.

In BLOR you may activate the option **2D pre-check**. This helps to find large errors in image co-ordinate files from an AATM.

5.2.3   Discussion of the Results

Let's begin with the results of the first modules, BLOR. Please open the file BLOR. LST and take a look at some of the text which may be more or less like the following:

```
NUMBER OF POINTS PER PHOTO
===============================
134   227   135   507   136   645   137   724   138   730
139   757   140   456   155   216   156   503   157   644
158   691   159   663   160   623   161   343   170   149
169   391   168   596   167   681   166   668   165   640
164   359
```

For each image of the block, the number of points is listed, ranging from 149 to 757 in this example.

Strip by strip, in the next section errors found by the programme are listed. The particular values may vary from those on your computer due to your manually measured tie points and other influences but will look similar to the following:

```
RELATIVE ORIENTATION MODEL  139 140  SIGMA0 = 103.12 μm
446 POINTS     BX = ..  BY = ..  BZ = ..
PHIR = ..  OMEGAR = .. GRADS  KAPPAR = ..


MODEL   139   140
POINT NO.  XL  YL  XR  YR  Y-PARALLAX   W   R   NABLA
139003     ..  ..  ..  ..     -1202.2  ..  ..       ..
139071     ..  ..  ..  ..      -579.8  ..  ..       ..
139084     ..  ..  ..  ..      -858.4  ..  ..       ..
139265     ..  ..  ..  ..       331.2  ..  ..       ..


RELATIVE ORIENTATION:
POINT  139003 DELETED FROM       139       140    ***
```

(To save space, only the point No. and the y parallaxes are listed here). Depending on the y parallaxes, a check was made to detect bad points. From all points in the model, containing in total 446 points, the worst ones are listed, and you can easily see that point No. 139003 is surely not correct. Therefore the programme suggests eliminating this point. The option **Error correction** which we activated in the parameters setting leads to an automatic elimination in the following BLUH modules (BLAPP, BLUH).

Near to the end of the list, you will see something like this:

```
ARRANGEMENT OF PHOTO NUMBERS, INTERNAL CAMERA NUMBERS
======================================================
STRIP  1:    134 1    135 1    136 1    137 1    138 1
             139 1    140 1
STRIP  2:    155 1    156 1    157 1    158 1    159 1
             160 1    161 1
STRIP  3:    164 1    165 1    166 1    167 1    168 1
             169 1    170 1


18 MODELS, MEAN NUMBER OF POINTS 354, MEAN SIGMA OF
REL. OR. 103.2
```

The programme sets up the strips (image number, internal camera number; the latter one all times equal 1, because all photos were taken with the same camera). In this example, the mean number of points per image is 354, the mean sigma naught (σ_0, standard deviation of unit weight) of the relative orientation is about 100 μm.

Further information is given, for instance the approximate values of the absolute orientation of each image, a photo number list and a final listing of the located blunders (errors).

Please close the file BLOR.LST and open the file BLAPP.LST, created from the second module. At the beginning, the error correction is listed, for instance:

```
134065              0       134       135     1
134066              0       134       135     1
136007              0       137       138     1
137388              0       137       138     1
137098              0       138       139     1
139003              0       139       140     1
139084              0       139       140     1
139071              0       139       140     1
. . .
```

This list was prepared by the automatic error correction in BLOR. The next listing reports whether the correction was successful:

```
134065 IN MODEL 134 135 REMOVED BY ERROR CORR. LIST
134066 IN MODEL 134 135 REMOVED BY ERROR CORR. LIST
136007 IN MODEL 137 138 REMOVED BY ERROR CORR. LIST
137388 IN MODEL 137 138 REMOVED BY ERROR CORR. LIST
137098 IN MODEL 138 139 REMOVED BY ERROR CORR. LIST
139003 IN MODEL 139 140 REMOVED BY ERROR CORR. LIST
139071 IN MODEL 139 140 REMOVED BY ERROR CORR. LIST
. . .
```

Again, a list of numbers of points per photo follows:

```
NUMBER OF POINTS PER PHOTO
================================
134   226   135   506   136   644   137   722   138   727
139   757   140   456   155   216   156   499   157   641
158   691   159   662   160   622   161   342   164   358
165   639   166   663   167   674   168   595   169   389
170   147

IN PHOTO  170   147 POINTS = LOWEST NUMBER
IN PHOTO  139   757 POINTS = HIGHEST NUMBER
```

As you can see, the amount of points differs slightly from the list above. This is due to the eliminated points in the error correction process. Nevertheless, we still have between 147 and 757 points per image—much more than necessary (9).

A similar listing follows, showing a statistic of the number of photos in which a single point was measured. For instance:

```
NUMBER OF PHOTOS/OBJECT POINT
PHOTOS/POINT     1     2     3      4     5      6
POINTS:          0   3183  1467    88     3      7
```

Most of the points are determined in only two (3183) or three (1467) neighbouring images. Points that occur in more than 3 images, in one single strip or in neighbouring strips, are not so many.

Now please close the file BLAPP.LST and open the file BLUH.LST, created from the last (main) module. After some statistical information and the list of control points you will see something like the following (here shortened a bit):

```
NO.ITER    MS CORR X     MS CORR Y     MS CORR Z    SIGMA 0
                                                    [microns]
=============================================================
0   .336577E+02   .678660E+01   .284580E+02      3600.6
1   .366986E+01   .330278E+01   .765927E+01        93.3
2   .317084E-01   .264608E-01   .930126E-01        93.2
```

The last column is of special interest, showing the σ_0, changing from iteration to iteration. As already mentioned before, this value is the standard deviation of weight unit and can be seen in case of bundle block adjustment as standard deviation of the accuracy of image co-ordinates. In the parameter setting (see above) we defined a maximum of 10 iterations, but, if no more significant improvement of σ_0 is reached, the process terminates.

The next two listings show the standard deviations of photo orientations and the photo orientations themselves, also stored in the file DAPOR.DAT.

Once again remember the scan resolution of 300 dpi or about 84.7 μm. A final result of more or less one pixel can be seen as sufficiently good. Nevertheless, this is only a standard deviation value, therefore let's take a look at the following results in the file. For the maximum of object points the remaining errors (residuals) are less than the limit for listing defined above (200 μm). But some few points show larger errors—this may look like the following:

```
168489    1141900.239   970948.637   1142.801      3
D.I.    168    429.1    -50.6    -37.2    -430.4  *
D.I.    167   -863.7      5.5     -2.2     863.8  ***
D.I.    166    429.7     20.7     29.6    -429.2  *
```

The amount of asterisks right-hand symbolises the size of the error. In our example you can see that point No. 168489 was found during the AATM in the images No. 168, 167 and 166. In image 167 there is a greater displacement in x, the values given in μm. As you know, we have many more points than necessary, so we will simply delete points with large errors completely. A bit more complicated is a situation like the next:

```
777770009   1137445.050   969699.794    1294.659        6
D.I.      134       67.3      362.9    -362.8      67.5  *
D.I.      155     1155.3     -208.1     205.9    1155.7  ****
D.I.      135      -12.1      298.3    -298.4      -9.9
D.I.      156     -386.9     -368.9     370.3    -385.6  *
D.I.      136     -224.8      369.7    -370.6    -223.5  *
D.I.      157     -574.2     -373.9     375.8    -572.9  **
```

As you can see from the point number as well as from the image numbers, this is a connection point (strip 1 and 2). These points are very important for the strip connection and should not be deleted in total if ever possible. From the residuals we can recognise that the only really bad point is in image 155, therefore we will only delete this measurement.

At the end of the listing, the final result of σ_0 is given:

OBSERVATIONS	UNKNOWNS	REDUNDANCE	SIGMA 0
			=======
22385	14370	8015	92.83
			[microns]

After this, the adjusted co-ordinates of all points are given, also stored in the file DAXYZ.DAT.

Now, how to do an error correction? Of course we can delete the respective lines directly in our object co-ordinate file AATM.DAT or in the export file DAPHO. DAT created from it, but this will be too much work. Remember that in the parameter setting we had an active option **Error correction** which makes the module BLOR to create an error correction file named DACOR.DAT. Please open this file which contains entries like this:

```
       136080         0        137      138      1
       138205         0        137      138      1
       134693         0        155      156      1
       136387         0        156      157      1
       136342         0        156      157      1
    777770006         0        156      157      1
    777770006         0        157      158      1
       137342         0        158      159      1
       160317         0        159      160      1
              . . .
```

The sequence of the entries is:

Old point number, new point number, left image, right image, activation flag (0 = deactive, 1 = active for a flexible handling).

You have the following options:

- If the new point number is zero, the point is deleted in this model (left and right image).
- If only the left or the right image number is greater than zero, the point is only deleted in that image.
- If both image numbers are zero, the point is deleted in the whole block.

With this information, you can edit this file using a simple ASCII editor, for example click onto the Text button right-hand in the main window. The only thing you must take into account is that the entries must be arranged in the way that the old point numbers (column 1) are in increasing order!

So, if we want to delete the two examples of bad points mentioned before using this file, we have to add the following two lines:

```
   168489              0          0        0    1
777770009              0        155        0    1
```

In this way, point 168489 is deleted completely, point 777770009 is deleted only in image 155. Attention: Before starting the block adjustment a second time, remember that as a standard the first module, BLOR (Pre 1), will overwrite our error correction file! To prevent this, don't run BLOR again but start with BLAPP (Pre 2)!

Now, the results of BLUH should be slightly better. If you like you can check the file BLUH.LST again and add more lines to the error correction list.

5.2.4 Improvement of Strip Connections

Up to now, the AATM have found tiepoints within each single strip, but the strips themselves are only connected by the control points and the manually measured connection points (see Sect. 5.1.7). With the results of the bundle-block adjust-ment (orientations and object co-ordinates) it is now possible to check for every object point in which images it may occur and also to calculate the approximate pixel co-ordinates.

Please start LISA FOTO again, select ATM > AATM, set all parameters as explained in Sect. 5.1.8 but now also activate the option data from BLUH, then click onto OK. The programme now measures connection points at the approximate positons, again first in pyramid images, then in the original images with full resolution. The results are stored in the file AATM_BLUH.DAT and transform them into the BLUH format DAPHO.DAT (image co-ordinates).

5.2.5   Second Run of BLUH

With the results from just before we process the bundle-block adjustment for a second time. Start BLUH_WIN again. Because we don't want to change any parameters we can just go to **Block adjustment > All/batch**. The BLUH modules BLOR (Pre 1), BLAPP (Pre 2) and BLUH (Main) are processed one after another and the final results are displayed afterwards.

Created files: BLOR.LST, BLAPP.LST, BLUH.LST.

5.2.6   Additional Analysis of the Results

After the block adjustment with BLUH is finished, we want to analyse the results and create an image, showing us for instance the positions of object and tie points. Use **Post processing > Analysis** or just click onto the respective button right-hand in the main window. Set the following parameters: Distance for neighbouring points 0.5 m, all others remain as before. After **OK** and a short time, the results are presented in an editor window. Let's again take a look at them.

In a first section, the control points are listed and the ranges of image and terrain co-ordinates are given. Then, neighbouring (and possibly identical) points are listed:

```
NEIGHBOURED OR IDENTICAL POINTS
===============================
```

		DX	DY	DZ	HOR DIST
134051	135035	.089	-.078	-.176	.118
134400	135270	-.095	.092	.387	.132
135232	136194	.049	-.077	-.439	.091
136416	158106	-.009	-.218	.332	.218
137076	138080	.405	-.218	-.123	.460
137362	138324	.072	.163	.107	.178
138187	139200	.379	.021	.105	.380
155309	156197	.000	.000	.000	.000
156454	157382	-.239	-.062	-.184	.247
160369	165080	.161	.249	-.319	.297
165114	166112	-.155	-.002	.033	.155
...					

The programme looks for points within a defined distance (separately in x/y and z) and lists all found point pairs. It is the user's turn to decide whether they are in fact identical.

A special comment should be given to points which appear in neighbouring strips like 136416 and 158106 (4th line). The horizontal distance is 0.218 m, and concerning the photo scale and the scan resolution leading to a pixel size of approximately 1.2 m in the aerial images, we can see that both points have a distance of less than one pixel. Therefore it is possible to unite the points, improving the strip connection in this way: Just run BLAPP (Pre2) and then BLUH (Main) again.

The next listing:

```
DIFFERENCES OF GROUND COORDINATES
===================================


CONTROL POINTS USED AS CHECK POINTS
ANALYSED DATA FILE: C:\STEREO\CAICE\BLUINF.DAT

    POINT     X    Y    Z PH/P    DX       DY       DZ       DS

    80001    ..   ..   ..    2  -2.066    .056    -.820    2.067
    80002    ..   ..   ..    6  -1.055   -.220    -.854    1.078
    80003    ..   ..   ..    4  -1.246    .077    -.671    1.248
    80004    ..   ..   ..    6   -.677    .830    1.025    1.071
    80005    ..   ..   ..    2   -.887   -.097   -1.008     .892
    80006    ..   ..   ..    5   1.357   -.857     .343    1.605
    80010    ..   ..   ..    5   1.785    .578     .331    1.876
    80011    ..   ..   ..    6    .169    .304    -.660     .348
    80012    ..   ..   ..    6  -1.074    .076     .403    1.077
    80013    ..   ..   ..    3    .723   -.949    1.603    1.193
    80014    ..   ..   ..    3   3.265   -.690   -2.201    3.337

SQUARE MEAN OF DIFFERENCES
SX = +/- 1.525   SY = +/-  .548   SZ = +/- 1.051   SS =
+/- 1.620
NX =        11   NY =        11   NZ =        11

MAXIMAL DIFFERENCES
MAX DX =  3.265  MAX DY =  -.949  MAX DZ = -2.201  MAX
DS =  3.337

SYSTEMATIC DIFFERENCES
SYSTX =     .027   SYSTY =     -.081   SYSTZ =     -.228
SQUARE MEAN OF DIFF. WITHOUT SYSTEMATIC DIFFERENCES
SX = +/-   1.524   SY = +/-     .542   SZ = +/-   1.026
```

(To save space, the values of x, y and z of the control points are not printedhere). As we can see, there are no extreme errors in the control point data.

```
    NUMBER OF PHOTOS / OBJECT POINT

    PHOTOS/POINT     2      3      4      5      6
    POINTS:       1690   1846     88    145    113
```

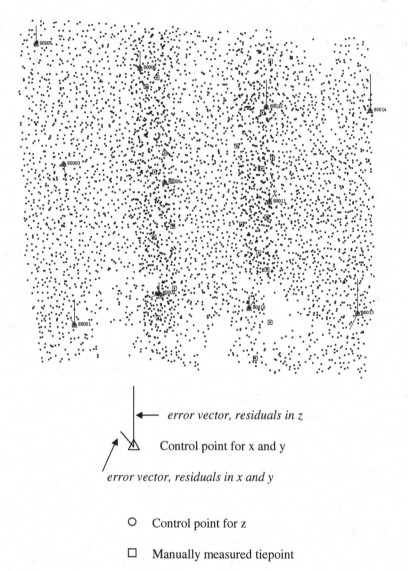

Fig. 5.8 Distribution of control- and tie points

In Sect. 5.2.3 we already discussed this information.

After closing the window, a vector graphics (format HP-GL) is created which shows all information we have selected: Image numbers and area covered by each image, control points and error vectors in x/y and z, and further all connection

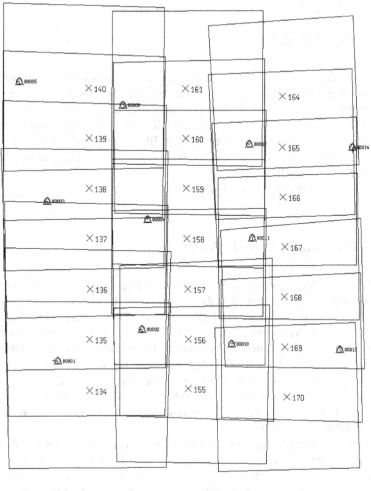

$\times$ 134   image number

area covered by the image

Fig. 5.9 Area covered by each image

points. The last ones are colour-coded to show the number of images in which they have been found (blue = 2, green = 3, cyan = 4, red = 5, magenta = 6 or more).

Figures 5.8 and 5.9 show two examples of the graphics output, the first for control and tie points (option **Photo location** in the **Graphics** section is set to "no"), the second for the representation of the areas covered by each image (option **Points** set to "no").

5.3   Mosaics of DTMs and Ortho Images

5.3.1   Model Definition

In the same way as known from our first example, for every model which we want to use in the following steps, a model definition must be carried out before. Go to **Pre programmes > Define model** and control/set the parameters: Activate **all models** (to make the model definition for all models in batch mode). For the exterior orientation, choose **Parameters from BLUH** and set the files to DAPOR. DAT (orientations) and DAXYZ.DAT (object co-ordinates). These are the results from BLUH as you will remember (Sect. 5.2.2).

Created files: 134135.MOD, 135136.MOD, …

5.3.2   Creation of a DTM Mosaic

Now the whole block is prepared for further processing—we have a large amount of object points/co-ordinates as well as the parameters of the exterior orientation of all images. In the same way as we did before it is possible to create DTMs and ortho images from each model, one after the other, and when the last model is processed we should be able to match the DTMs and also the ortho images together to mosaics.

But, as you already have seen in some examples before, it is nice if we can do the work automatically model by model in a batch mode, and this is also possible here. Let's start with the creation of all DTMs and finally put them together into a mosaic. But beware—also this is a time-consuming process, so it is a good idea to prepare another cup of coffee.

Start **Processing > Stereo correlation**, activate the option **All models + create mosaic**, maintain all other parameters. The **Object co-ordinates** we now take from the file DAXYZ.DAT, then click onto **OK**. Model by model, the programme will create a DTM file with a name like GT_<left image, right image>. IMA. Finally, all these files are matched together to the output file GITT.IMA. In between, an info window informs you about the progress of correlation.

After the programme has terminated, the result will be displayed (see also Fig. 5.10). Created file: GITT.IMA.

5.3.3   Creation of an Ortho Image Mosaic

Similar to the chapter before, we will create an ortho image mosaic automatically. This has not only the advantage of faster work but gives us also the possibility to adjust the grey values of the input images to get a final ortho image with (nearly) no visible grey value edges (see this effect also in Sect. 4.7.3).

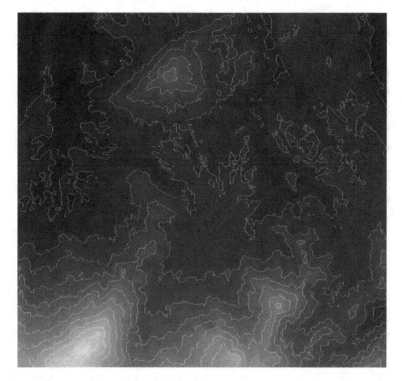

Fig. 5.10 DTM mosaic, 25 m contours overlaid

Go to **Processing > Ortho image,** choose the option **All images** and let the **Colour value adjustment** be activated. File names: Terrain model GITT.IMA from our last chapter, output image ORTHO.IMA. Again, an info window informs you about the progress of work. After the programme has finished, the result will be displayed (Fig. 5.11).

Remark: The ortho image mosaic as well as the DTM mosaic from before is of course geocoded. To use such images for commercial GIS software you have to convert them into a standard format like TIF. For this, select **File > Save as** in the display window. Beside of the image file, a so-called *worldfile* is created, containing the geometric parameters which can be used in the GIS software.

Created file: ORTHO.IMA.

5.3.4   Shaded Relief

Let's play a bit with the various possibilities of DTMs and image combination: As an idea, we want to calculate a shaded relief image and combine this with our ortho image mosaic to produce a bit more spatial impression.

Exit the LISA FOTO programme and start LISA BASIC which we will use for the rest of this tutorial. Then, carry out the following steps:

Fig. 5.11 Ortho image mosaic

- Load our DTM GITT.IMA into the raster image display, then select **DTM >
  Shading**, then OK. Now select **File > Save as** and use SHADE.IMA as name
  for the output image. Remark: The default values **Direction** 315° and
  Inclination 45° For the "light source" refer to the shading principle in cartog-
  raphy, "light from top left".
- Close the image display and go to **Image processing > Matching > Addition.**
  Then **Weighted**, set the **Weight** for image 1–70 %, **Image 1** ORTHO.IMA,
  Image 2 SHADE.IMA, **Output image** ADDI.IMA, then **OK.**

The result shows a combined image in the way that each grey value is calculated
by 70 % from the ortho image and 30 % from the shaded relief image.
  Created files: SHADE.IMA, ADDI.IMA.

5.3.5   *Contour Lines Overlay*

Similar to Sect. 4.7.4, you may calculate contours from the DTM mosaic, useful for
an overlay over the ortho image mosaic. As we already discussed in that chapter, it
is a good idea to filter the DTM before, giving smoother contours as a result.

Fig. 5.12 Ortho image mosaic draped over the DTM mosaic (see also the book cover)

Go to **Terrain models > Filtering**. Maintain all parameters, set the input file to GITT.IMA, the output file to GITT_FLT.IMA, then **OK**.

Now go to **Terrain models > Contours vector**. Set the parameters **Equidistance** to 25 m and **Tolerance** to 0.5 m, define CONTOUR.DAT as the name of the output file, then click onto **OK**. The result will be displayed.

Close that window and open the raster image display again, load ADDI.IMA, then select **Overlay > Vector graphics** and load the file CONTOUR.DAT. Now use **File > Save as** and store the result as ADDI_2.IMA.

Created files: GITT_FLT.IMA, CONTOUR.DAT, ADDI_2.IMA.

5.3.6 3D View

As a final graphics result we would like to calculate a 3D view of our complete area. Remember that we have *height information* (our DTM mosaic) and *surface information* (for instance our ortho image mosaic combined with 30 % shading). From these two "layers" it is possible to create a 3D image from any viewing direction. This is the way.

Start the raster image display and load for instance ORTHO.IMA (or ADDI.IMA or ADDI_2.IMA, as you like). Select **View > 3D view**, then the option **actual image** and the DTM GITT_FLT.IMA, then **OK**. In the tool window at the lower-right part of the display you now can change the values for azimuth, inclination, exaggeration and zoom. If you now close the tool window you can store the result for instance as VIEW_3D.IMA (see Fig. 5.12).

Created file: VIEW_3D.IMA.

Chapter 6
A Soil Erosion Experiment

From now on we will use images from digital cameras and will see how photogrammetry can be used in "close-range" applications. In this 3rd tutorial, the situation differs from aerial photogrammetry only in relation to the dimensions of the "flying height" and the base.

6.1 The Situation

To get detailed information about soil erosion, an artificial test field was constructed and the surface photographed before and after several "rainfall" events. Figure 6.1 shows the test field with control points at the borders and the camera position. If this method of modelling is of interest for you, see for instance Wegmann et al. (2001), Rieke-Zapp et al. (2001) and Santel (2001).

The tests were carried out in collaboration between the Institute of Photogrammetry and GeoInformation (IPI), University of Hannover, and the National Soil Erosion Research Laboratory (USDA-ARS-NSERL), West Lafayette, Indiana, USA. Thanks to both organisations for the data!

The images were taken with a digital monochrome camera, type Kodak DCS 1 m with a Leica Elmarit R 2, 8/19 mm lens. Some words about the object co-ordinate system: Of course it makes no sense to use a system like Gauss-Krueger or UTM—therefore, in cases like this we will use a Cartesian local system, sometimes called "non-world" in textbooks. But—what is non-world? Lunar? So, let's better say *local*.

From the complete data set we will use one stereo model showing the initial situation before rainfall and another stereo model of the same region after 4 rainfall events, showing erosion on the whole area as well as a linear runoff (drainage) system. From both cases, we will create a DTM, then calculate a differential DTM afterwards to evaluate the amount of eroded soil.

© Springer-Verlag Berlin Heidelberg 2016
W. Linder, *Digital Photogrammetry*, DOI 10.1007/978-3-662-50463-5_6

Control points   Camera

Soil surface

Fig. 6.1 Test field for soil erosion, a camera position, control points. From Santel (2001)

6.2   Interior and Exterior Orientation

Fiducial marks are only used to establish the interior orientation for photos taken
from a traditional metric film camera. Digital cameras applied with an area CCD
sensor do not need them because each CCD element gives the same image pixel
every time.

Therefore, the method of camera definition differs a bit from that you already
know, and the interior orientation is given directly and must not be carried out
image per image. Start LISA FOTO with the project TUTOR_3.PRJ, then go to **Pre
programmes > Camera definition > Digital,** and key in the following parameters: **Pixels in x** 2036, **pixels in y** 3060, **pixel size** 9 μm and **focal length**
18.842 mm. These data you can get usually from the camera's manual. Now click
onto the button **calibration data,** select the option **from file (R, DR)** and the file
RADIAL_VZ.DAT. Set the values of the **principal point** to −0.28 each, then **OK.**
As name for the output file choose KODAK_DCS.CMR. We will learn more about
camera calibration in Chap. 8.

After a click onto **OK,** in fact *two* files are created: KODAK_DCS.CMR, the
camera definition, and KODAK_DCS.INN containing the interior orientation data
for all images.

The next step will be the exterior orientation, similar to our first tutorial
(Sect. 4.2.5). The control points near the upper and lower photo borders are
signalised and labelled, their co-ordinates stored in the file CONTROL.DAT.
If you like you can make the orientation work for both images of both situations

(No. 1005 left, 1004 right before rain, 5005 left, 5004 right after rain) now, use Sect. 4.2.5 for advices. If not, simply use the files 1004.ABS, 1005.ABS, 5004. ABS and 5005.ABS from the Springer server (...\tutorial_3\output).

The control points and their co-ordinates (all values in [mm]):

Point No.	X	Y	Z
2003	3806.904	2828.597	1095.195
2004	3369.888	2829.467	1273.887
2005	3021.855	2825.091	1274.496
2006	2549.527	2819.946	1380.263
2008	1875.663	2803.879	1530.468
2023	3882.687	878.742	1142.570
2024	3546.677	871.331	1202.315
2025	3144.343	865.521	1266.954
2026	2693.791	856.243	1328.317
2027	2253.322	852.145	1410.972
2028	1822.189	844.232	1526.819

Created files: KODAK_DCS.CMR, KODAK_DCS.INN, 1004.ABS, 1005. ABS, 5004.ABS, 5005.ABS

6.3 Model Definition, Start Points

As usual, our next step is to define the model. Go to **Pre programmes > Define model** and set the following parameters: Left image 1005, right image 1004, **parallax correction** 3 pixels. Use exterior orientations from ABS files, as object co-ordinates file take CONTROL.DAT. Go on with **OK**.

Start **Processing > Stereo measurement**, start height is 1200 [mm]. Now try to set homologous points by adjustment of the z value with the central mouse wheel in several positions, near the model edges, in the model centre etc. You will find that in some positions the y parallaxes will reach two or three pixels—this may give a negative influence for example if we want to generate a DTM by image matching, therefore we've set the option **parallax correction** to 3 pixels in the model definition.

Try to find points with a good contrast in all directions. A file with such points is already prepared for you (START_1000.DAT). For instance, you can use the **Processing > Stereo measurement** option and then **Overlay > Vector data**, then select this file to see what we mean. But of course you may measure points for yourself using the **Digitise > Points/lines** option (see Sect. 4.6.3 for instance) (Fig. 6.2).

As a result, we have a file with a sufficient number of points to improve the model definition. Start **Pre programmes > Define model** again and set the option

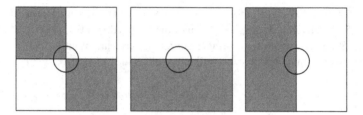

Fig. 6.2 Schematic drafts of points with good contrast. *Left* suitable for all purposes. *Middle* suitable only for y parallax correction. *Right* suitable only for measurement of the x parallax (→ height or z value)

parallax correction to 3 pixels. Contrary to the first start of this option, set the file Object co-ordinates to START_1000.DAT, then click onto OK again.

In the same way we have to prepare the model "after rainfall", the images 5005 (left) and 5004 (right). The point file START_5000.DAT is also prepared for you, but again and for training you may measure own points. After that carry out the model definition with the same parameters like used before.

Created files: START_1000.DAT, START_5000.DAT, 10051004.MOD, 50055004.MOD

6.4   DTM Creation

First go to Pre programmes > Select model and choose 10051004. Then start Processing > Stereo correlation and set the following parameters: px ± 8 pxls, Correlation coefficient r > 0.7, window 7 by 7 pxl, 3 iterations. Set the name of the output image (DTM) to GITT_1000.IMA, maintain all other parameters as set by default, then click onto OK. If necessary, see Sect. 4.6.2 for further information.

Now go to Pre programmes > Select model and choose 50055004. Start Processing > Stereo correlation again, use the same parameters like before, set the name of the output image (DTM) to GITT_5000.IMA, then click onto OK.

Some remarks concerning the parameters we have used: As you can see, the images 1004 and 1005 have low contrast in some areas. Therefore we selected relatively large windows (13 by 13 pixels) to get a better statistical base, and a high number of iterations (10). With this we can improve the result—in particular, we get a high number of correlated points, remaining only few positions filled by interpolation. In the second model (after rainfall) the contrast is better due to the relief. Therefore the windows can be smaller, and besides there is a rule "the more relief, the smaller the correlation windows" (see also Sect. 4.6.1).

Created files: GITT_1000.IMA, GITT_5000.IMA

With the just created DTMs of both situations we are able to calculate a differential DTM, showing us the effect of erosion and giving us the possibility to calculate the amount of soil washed out during four rainfall events.

6.5 Differential DTM

Please close LISA FOTO and start LISA BASIC, use TUTOR_3.PRJ like before. Go to Terrain models > Matching > Differential DTM and set the first file name to GITT_1000.IMA, the second to GITT_5000.IMA, keep the name of the output image to DIFF.IMA and maintain all other parameters, then click onto OK. The next window informs you about the minimum and the maximum value of height difference. Just click onto OK again and display the result DIFF.IMA, if you like.

Both DTMs as well as the differential DTM are shown in Fig. 6.3. For a better representation of the terrain surface, 10 mm contours were created and overlaid (see Sect. 4.7.4 how to do this).

Now, as a last step in this example, let's calculate the amount of soil washed out: Use Terrain models > Load/change DTM to control whether the differential DTM (DIFF.IMA) is our actual one. Then start Terrain models > Numerical evaluation > Volume differences, the name of the output file keep as STAT. TXT. The result will look more or less like following:

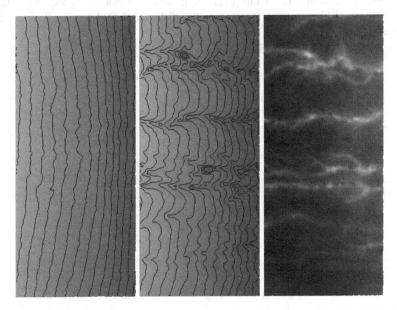

Fig. 6.3 Situation before rain (*left*) and afterwards (*middle*), 10 mm contours overlaid in both images, differential DTM (*right*)

```
Volume differences:
Decrease          21992615.0285 mm³
Increase                 0.0000 mm³
Saldo            -21992615.0285 mm³
in average              19.5560 mm height change.

Resolution:              3.0000 mm in xy,
resp.                    0.0038 mm in z.
```

The results are given in [mm³] and [mm] according to the length unit selected in the project definition. Within our model area about 21.99 dm³ of soil was eroded with an average height change of 19.6 mm between the terrain models.

Finally, let's talk about some problems we can find in the relief "after rain": As you may have seen, the valleys have very steep slopes in some regions, caused by heavy erosion of the soil which is not protected by vegetation. This effect is known in reality as "gully erosion". As a consequence, we have hidden areas in some parts (see also Fig. 4.15). This in conjunction with the dark, nearly contrast-free bottom of the valleys may lead to problems in the matching process and the derived DTM, for example unrealistic holes and peaks within the valleys. Such incorrect DTM heights have of course an influence on the differential DTM as well as on the volume differences calculated from it.

Created files: DIFF.IMA, STAT.TXT

Chapter 7
Wave Measurements

7.1 The Situation

The last example already belongs to the so-called *close-range photogrammetry* but in fact, it has a geometric situation similar to the aerial case (vertical images). The next example shows a more typical close-range or terrestrial case.

For coastal protection, it is necessary to know as much as possible about wave movement and wave energy. Therefore, stereo images of waves rolling onto the shore were taken from two buildings situated in Norderney, an East-Friesian island in northern Germany, using four digital cameras, type Ikegami SKC-131 with a Cosmicar/Pentax 12.5 mm lens (wide angle). Figure 7.1 gives you an impression about the area and the camera positions. For our example we will use images from cameras I and II.

The interdisciplinary project "WaveScan" was carried out by the Institute of Fluid Mechanics (ISEB) and the Institute of Photogrammetry and GeoInformation (IPI), both University of Hannover, and sponsored by the Federal Ministry of Education and Research (BMBF), code 03KIS026. All rights of the image data are owned by the IPI.

The cameras were activated simultaneously in time intervals of 1/8 s by a wireless equipment, developed by Dr.-Ing. D. Pape. This is necessary because the object (water surface) is moving. The differences between this and all previous examples concerning the geometric situation are:

- The images were not taken "camera looking down" giving us vertical images, therefore, we have oblique images with a large variation of the scale. As a result, the values of the rotation angles φ and ω are no more near zero.
- The cameras were situated on top of a building of only 45 m height, our field of interest has an extension of some 100 m in front. This leads to several hidden areas, the backward sides of greater waves.
- The projection rays [projection centre $\rightarrow$ image centre] are not parallel but slightly convergent.

© Springer-Verlag Berlin Heidelberg 2016
W. Linder, *Digital Photogrammetry*, DOI 10.1007/978-3-662-50463-5_7

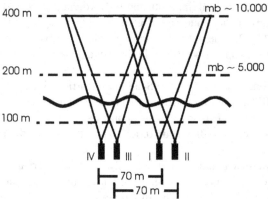

Fig. 7.1 The test area (*above*) and the camera positions on top of two houses (*below*). From Santel et al. (2002)

As you will see, there are further problems: The images are not very sharp as a result from the misty weather and a resolution of only 1296 by 1031 pixels. And, the rolling waves produced linear parallel forms in the images which lead to the effect of repetitive structures, already discussed in Sect. 4.6.1.

The goal is to calculate a "DTM" of the water surface. In principle, this is nothing new for us, and therefore we will only take a look at the differences in the work flow, and what it means in particular to the exterior orientation.

From the complete data set collected in the project, a sub sequence is prepared for this example with the images 100001 ... 100005 (left) and 200001 ... 200005 (right). The images were taken with camera I and II from the right building (see Fig. 7.1).

7.2 Interior and Exterior Orientation

Start LISA FOTO using the project TUTOR_4.PRJ, then go to **Pre programmes > Camera definition > Digital**. With the file open button you can see that two cameras are already prepared: CAMERA_1.CMR and CAMERA_2.CMR, both calibrated which means that we have values for the principal point and the radial-symmetric lens distortion for both cameras (see Chap. 8). The image names (6 digits) serve to connect them to the respective camera: 100001 ... 100005 → camera 1, 200001 ... 200005 → camera 2 (see Sect. 10.7.2).

Remember Sect. 6.2: The interior orientations for all images taken with these cameras are also defined now.

The control points we will use for the exterior orientation are shown in the Figs. 7.2 and 7.3, their co-ordinates are listed below and prepared in the file CONTROL.DAT.

Point No.	X	Y	Z
100	2575400.404	5953951.649	9.008
101	2575400.130	5953951.787	4.435
104	2575400.128	5953951.799	2.634
105	2575431.675	5953971.315	7.275
106	2575431.682	5953971.309	4.443
109	2575431.683	5953971.298	2.630
110	2575406.817	5953976.222	11.862
111	2575310.374	5954111.996	8.712
144	2575489.288	5953866.850	7.414
145	2575490.007	5953867.365	7.417
146	2575514.772	5953884.686	7.433
147	2575518.685	5953887.285	7.423
153	2575490.736	5953867.634	10.188
154	2575512.349	5953882.868	10.197

Like in our example before, we must carry out the exterior orientation by measuring the control points manually. Go to **Pre programmes > Orientation measurement**, load image No. 100001, and choose **Measure > Exterior**

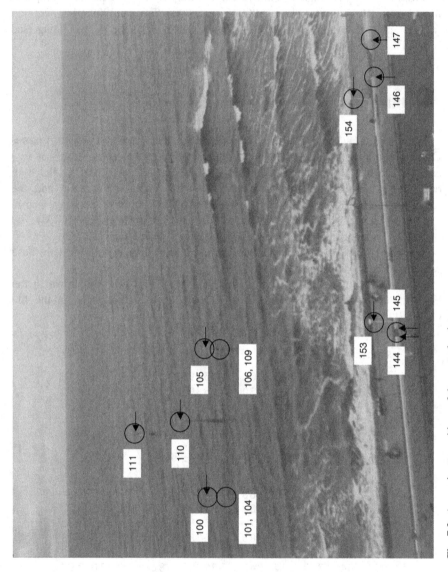

Fig. 7.2 Approximate positions of the control points

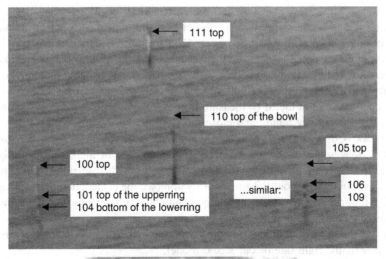

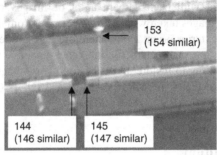

Fig. 7.3 Positions of the control points in detail

orientation. Select CONTROL.DAT as the name of the control point file. Now, here comes the first difference with our examples before: Due to the fact that we have no vertical images, it may happen that the orientation process will have some difficulties to converge. In this case is advisable to select the first three or four points in the way that they cover a maximum part of the image. In our example these may be the point sequence 110, 144, 147 and then 154.

Alternatively you may enter the (approximate) co-ordinates of the projection centre in the input window. This will also help the programme to calculate the parameters of the exterior orientation:

Image No.	X	Y	Z
100001 ... 100005 (left):	2575527.7	5953793.9	45.3
200001 ... 200005 (right):	2575541.6	5953807.3	45.3

Now use Figs. 7.2 and 7.3 to find the correct GCP positions—this may cause some problems because the images are not very sharp, see above. But, try to do your best, it's a good training! See Sect. 4.2.5 for more details if necessary.

After the last point is measured, take a look at the results below (residuals, standard deviation). If they are bad, you may mark the worst point in the list window and click onto the (De)activate button. The point is now marked with an S (= skipped) instead of the M (= measured). If you are satisfied, save the orientation data using the Ready button and close the window.

Remark: Concerning the results, always remember the pixel size. In case of scanned images this means the scan resolution, in case of images from a digital camera (like here) it means the resolution of the sensor. The camera used here has a sensor with pixels of about 6.7 μm length/width (see the camera definition). Even if we take into account the situation (oblique and not very sharp images) the standard deviation should not be more then 3 or 4 pixels.

In the same way like before carry out the exterior orientation for image No. 200001, the right one of our stereo model.

Now remember that with each camera a lot of images were taken in time intervals of 1/8 s. All images taken with the *same* camera from the *same* position of course have identical exterior orientations, therefore it is not necessary to measure (calculate) the orientation parameters for the other images of our sequence. In Sect. 7.5 you will see how to handle all of the following images automatically.

Created files: CAMERA_1.CMR, CAMERA_1.INN, CAMERA_2.CMR, CAMERA_2.INN, 100001.ABS, 200001. ABS.

7.3   Model Definition

As already pointed out in Sect. 6.3, we can help the programme to generate DTMs by the measurement of several well-distributed start points. For this example, a file with manually measured points (START_PNT.DAT) is prepared and may be used here.

Select Pre programmes > Define model and set the following parameters: Left image 100001, right image 200001, parallax correction 1 pxl. For the exterior orientations choose Parameters from ABS files and set the name of the object co-ordinates file to START_PNT.DAT, then click onto OK.

Created file: 100001200001.MOD.

7.4   DTM Creation

First go to Pre programmes > Select model and choose 100001200001. Then start Processing > Stereo correlation. Before going on, let's remember the difficulties of the situation, low contrast and repetitive structures. To handle them we will use the manually measured points as start points, a low z range and a high correlation coefficient threshold value.

Set the following parameters: px +/−3 pxl, correlation coefficient r > 0.7, correlation window 11 by 11 pxl, number of iterations 3.

Object co-ordinates START_PNT.DAT. This file includes also a border polygon. If you want to measure such a polygon by yourself, use the stereo measurement module and select **Register > Points/lines**, there choosing the code **Free cut area polygons**, and after that setting some **Delete start points**. See Sect. 10.7. 18 and the appendix, part 1, for further details. Set the name of the output image (DTM) to GITT_TST.IMA and activate the option **Mean filter** 5 × 5 pixels. If necessary, see Sect. 4.6.2 for further information.

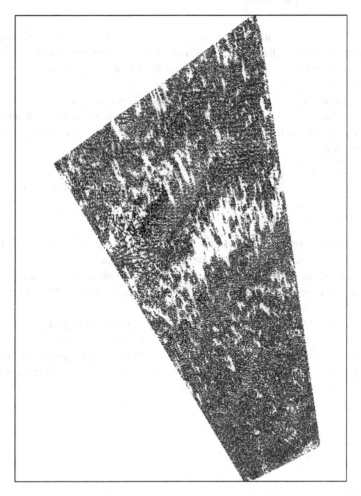

Fig. 7.4 Points found by correlation, showing the wave structures. The cameras are looking from bottom right

After OK, the matching process already known from previous examples begins. When the DTM improvement is calculated, you can see the wave structure displayed on the screen, looking somewhat like Fig. 7.4.

Please recognise that the points found by correlation are concentrated in some areas, showing front and top of the waves. Large areas without points are located on the backward slopes of the waves, parts of them hidden in the images as a result of the relatively low height of the camera positions, other parts with very low contrast.

Created files: GITT_TST.IMA.

7.5 Image Sequences

In this chapter we want to see how sequences of stereo pairs (time series) can be handled automatically. Remember that we have fixed camera positions—interior and exterior orientations as well as the model definitions (y parallax correction) can be seen to be constant.

The main idea is that, depending on the time interval Δt, changes in the surface model between t_i and $t_i + \Delta t$ are not too large. Therefore, if the time interval is small enough (depending on the speed of changes/movement), we can use the following strategy:

- Prepare the first model and calculate the first surface (DTM)
- Extract a set of well distributed points from this surface, serving as start points for the following model
- Calculate the next surface with these start points

As an option, the generated start points of all models can be stored. Also, an ortho image may be produced from every surface model.

In the following we will use five models from the whole data set: Images 100001 ... 100005 (left) and 200001 ... 200005 (right). Please select Processing > Image sequence and set the following parameters.

First model, image left 100001, right 200001; Last model, image left 100005, right 200005. Activate the option create ortho images, then OK.

The next window you know from the stereo correlation in the previous chapter—please take the same parameters like there. It is not necessary to define the name of the output image (the surface model) because all names are generated automatically.

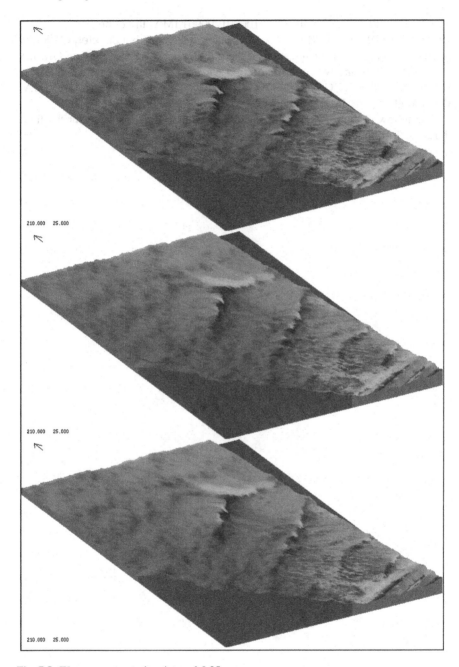

Fig. 7.5 Wave movement, time interval 0.25 s

The DTMs will get names like GT_100001200001.IMA, the ortho images names like OR_100001200001.IMA and so on. Make sure that the option **Interpolation** is activated, necessary for the creation of the ortho images. After a click on OK, the programme starts and will inform you in a window about the progress.

Created   files: GT_100001200001.IMA   ...   GT_100005200005.IMA, OR_100001200001.IMA ... OR_100005200005.IMA.

The following figure shows the results (ortho images in 3D-view) for the first, third and fifth model (Fig. 7.5).

Chapter 8
Lens Distortion and Calibration

8.1 Introduction

In this tutorial we want to show how easy it is to use photogrammetry with images taken by your own with a standard consumer camera. Of course, the optical and mechanical quality of such cameras is clearly less than those of special metric ones. Therefore we will not expect very high accuracy of the results, and furthermore we have to take a look at the image distortions and discuss a simple way to minimise them.

Most of these kinds of cameras are equipped with a zoom lens, the focal length can be set in a large range between wide angle and telephoto. A typical effect in wide angle mode are the barrel-shaped distortions, that means, straight lines near the image borders are shown bended to the borders. This effect usually will be less or zero in medium focal lengths and may turn into the opposite form (pincushion-shaped) at telephoto mode (see Figs. 8.1 and 8.2).

Beside these so-called radial-symmetric distortions which have their maximum at the image borders there are more systematic effects (affine, shrinking) and also non-systematic displacements. The distortions depend among others on the focal length and the focus. To minimise the resulting geometric errors efforts have been undertaken to find suitable mathematical models (for instance see Brown 1971).

In most cases the radial-symmetric part has the largest effect of all. The resulting errors are symmetric to the so-called *principal point*. This is the point in which a projection ray is normal (rectangular in all directions) to the film plane or the CCD sensor. In the camera definition module of LISA FOTO you can activate the option calibration data and then, if you know the parameters of one of the usual models, you can enter them there. See Sect. 10.7.3 for more details.

© Springer-Verlag Berlin Heidelberg 2016
W. Linder, *Digital Photogrammetry*, DOI 10.1007/978-3-662-50463-5_8

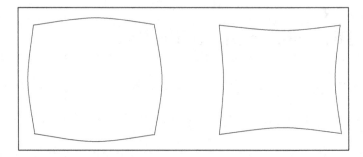

Fig. 8.1 Barrel-shaped (*left*) and pincushion-shaped (*right*) distortions

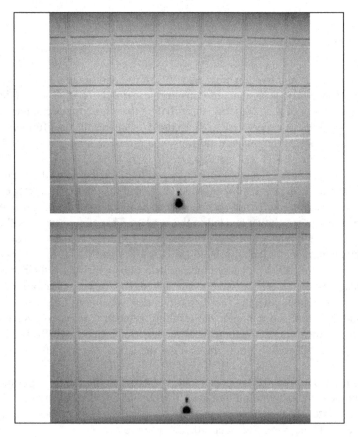

Fig. 8.2 Effects of lens distortion. *Above* wide angle, barrel-shaped distortions. *Below* normal angle, very few distortions

8.2 Lens Calibration

To capture distortion values we will use a calibration pattern, an example is prepared for you: Display the file 2888.JPG on a large flat-screen monitor in full-screen mode.

In this example, the pattern is given by 88 target marks, arranged in 8 lines à 11 marks. If you like to create your own calibration pattern, please note the following: Arrange 10 ... 20 unique marks in 5 ... 10 lines with equal distances in x- and y-direction on a plate or a wall. The overall size should correspond with the size of the objects you want to measure later. Now select the focal length you want to use later, then take a photo from the pattern in the way that all of your marks are inside of the image like in Fig. 8.3 an save it with a numerical name, for instance 1000. JPG.

In our example given here, a LUMIX DMC FZ28 camera from Panasonic Corporation was used with 4.8 mm focal length. Start LISA FOTO using the project TUTOR_5.PRJ, then go to **Pre programmes > Orientation measurement** and load the file 2888.JPG, after that select **Measure > Calibration pattern**. You have to define the number of targets in x (= 11) and y (= 8), then click **OK**. The image is now pre-positioned near to the lower-left target mark. In the same way you already know from the orientation measurements in the previous tutorials, move the target with the middle mouse button pressed "under" the measurement mark, then click the left mouse button. Move the mouse a little bit—then measure the next target (lower-right), flowing by the targets located upper-right and upper-left in the

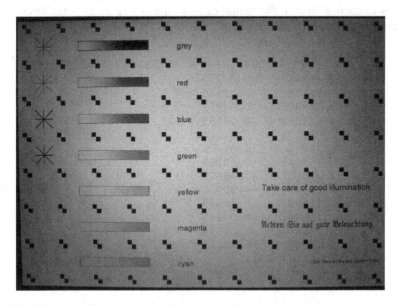

Fig. 8.3 Photo of a calibration pattern, displayed on a flat-screen monitor

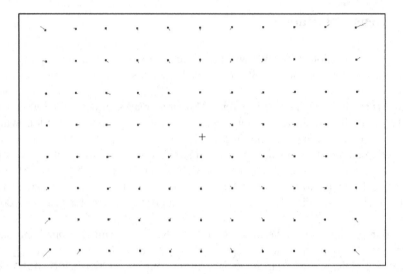

Fig. 8.4 Distortion vectors at the target marks and position of principal point (*cross*)

image. After the fourth target is measured, all others are processed automatically. Finally all residuals and the standard deviations are shown in the listing below. Click onto the **Ready** button, then close the measurement window and start the raster image display. A graphics called CALIB_2.IMA is displayed (Fig. 8.4):

At every point (target) you can see an error vector representing the size and direction of the distortions. From all of these vectors the position of the principal point is calculated. Within the central camera directory (c:\users\public\lisa\cmr) a file called LUMIX.CAL was created:

```
-0.00057160 -0.00091317 -0.00769563  0.00287713
       0.04386       0.08773
       3.57175       0.03570
       3.58990       0.02233
       3.56905       0.03951
       3.58110       0.02219
       3.14011       0.01679
       2.75184       0.00153
       2.42569      -0.00349
       2.20596      -0.00975
       2.12269      -0.00999
       2.20076      -0.01245
       2.41648      -0.00710
       2.74337      -0.00218
        . . .          . . .
```

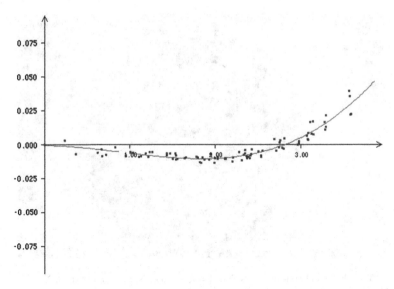

Fig. 8.5 Polynomial modelling the radial-symmetric displacements at the 88 target marks

Second ... last line: Distance from the principal point and displacement for each measured target mark in [mm]. First line: Coefficients of a 3rd-order polynomial modelling these values.

Now start **Pre Programmes > Camera Definition > Digital,** load the camera file LUMIX.CMR and click onto **calibration data.** You will find the second formula activated and the values from the first line of the file LUMIX.CAL entered as coefficients K1 ... K4. Also the values of the principal point are given. Click onto OK and start the raster image display (Fig. 8.5).

Within the camera definition a graphics called CALIB_1.IMA was created, showing the radial-symmetric distortions of the lens.

Created files: CALIB_1.IMA, CALIB_2.IMA.

8.3   A Simple Self-made Model

To encourage you for own experiments with your camera, we present a very simple example. Two photos were made from a tin plate with a decorative relief inside. A paper sheet of A4 size (297 × 210 mm) was placed under the plate and the corners from the sheet were used as control points. So, we have only 4 control points with equal height—not really good, but even this is possible (Fig. 8.6).

The file CONTROL.DAT is prepared with the co-ordinates of the four points. As you already know, the next steps are: Measurement of the exterior orientation in both images (489.JPG and 490.JPG), then the model definition (see Sect. 4.3 for instance), after this you can start the stereo measurement.

Fig. 8.6 Test object and control points: No. 1 = *lower left*, No. 2 = *lower right*, No. 3 = *upper right*, No. 4 = *upper left corner*

Created files: 489.ABS, 490.ABS, 489490.MOD.

To give a further example for the various possibilities of photogrammetric evaluation, the following steps were carried out:

- Measurement of some start points for stereo correlation (DTM creation) plus a surrounding border polygon of the central relief of the plate
- DTM creation with interpolation
- In LISA BASIC: File > Export > DAT/vector, input file was the DTM, then Grid and option all points. The output file is a 3D point cloud, here loaded into the vector display and shown in 3D representation (Fig. 8.7):

Fig. 8.7 3D point cloud of the relief

Chapter 9
Two Final Examples

9.1 Stereo Images from Satellites

Since few years a new chapter of photogrammetry has opened: Stereo images taken from satellites. If you remember the examples from this book we have seen that photogrammetric methods in principle do not depend on the size of the object or the distance between camera positions and object: We can handle aerial images in more or less the same way than terrestrial images in close-range applications. It is also possible to use images taken with the help of stereo-microscopes. Therefore we should be able to handle stereo pairs of images of any scale or ground resolution—why not satellite images?

We have to deal with one significant difference, the camera geometry. In all of our examples we used images from central-perspective cameras, as you remember: For each image we have one projection centre, the intersection point of all projection rays. The central perspective leads to radial-symmetric displacements according to the relief (for instance, see Fig. 4.15), and these are pre-requisite that we see the images stereoscopically and that we can measure three-dimensional object co-ordinates.

The digital cameras which are operated on satellites usually have a quite different geometry. In contrary to the central-perspective ones we know (CCD area sensor, "frame" camera) they only have a line sensor, "looking down to earth" and scanning the earth's surface line by line. As a result we have central perspective only within a single line (across the flight direction) but a parallel projection from line to line (in flight direction).

Modern satellite cameras often have more than one line: For instance, one line is straight looking down, one is looking backward, another forward. In this way stereo images can be obtained: One image from the "looking forward" scan line, the other some minutes later from the "looking backward" scan line (Fig. 9.1).

Due to the geometry just described, we cannot use our standard approach from the examples before. Remember that we calculated image co-ordinates from object

© Springer-Verlag Berlin Heidelberg 2016

W. Linder, *Digital Photogrammetry*, DOI 10.1007/978-3-662-50463-5_9

Fig. 9.1 Geometry of stereo
images from satellites.
From Jacobsen, 2007

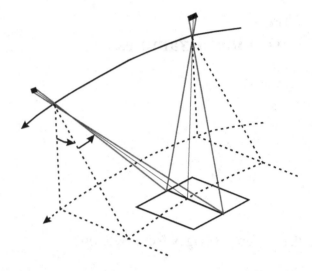

co-ordinates (x, y, z) using the collinearity equations (formula 4.3.1). In the cases
handled here, these equations must be replaced by different ones, using the
so-called rational polynomial coefficients (RPCs; see for instance Grodecki 2001):

$$x' = \frac{P_1(x, y, z)}{P_2(x, y, z)} \qquad y' = \frac{P_3(x, y, z)}{P_4(x, y, z)}$$

with $P_n(x, y, z) = a_1 + a_2*y + a_3*x + a_4*z + a_5*y*x + a_6*y*z + a_7*x*z + a_8*y^2 + a_9*x^2 + a_{10}*z^2 + a_{11}*y*x*z + a_{12}*y^3 + a_{13}*y*x^2 + a_{14}*y*z^2 + a_{15}*y^2*x + a_{16}*x^3 + a_{17}*x*z^2 + a_{18}*y^2*z + a_{19}*x^2*z + a_{20}*z^3$

(9.1.1)

As you can see, in total 80 coefficients a_i are used here—each 20 for the
polynomials $P_1 \ldots P_4$. These coefficients are delivered together with the image data
and give an approximation of the exterior orientation. Depending on the satellite
(for instance Cartosat-1, Ikonos, Quickbird, OrbView-3) we will use one or more
ground control points to improve the orientation (see Jacobsen 2006, 2007).

After the orientation is carried out for every image the user can go on in the same
way like in "traditional" photogrammetry: Object co-ordinate measurements, image
matching to obtain a DTM, creating of ortho images and so on. The ground res-
olutions of actual satellite data as well as the elevation accuracy are high. For
instance, data from the Cartosat-1 satellite have a ground resolution of about 2.5 m

(panchromatic), the RMS in z can reach a value of about 2 m as several test have shown.

Stereo satellite images together with RPCs can be processed with the LISA programme FFSAT which was copied onto your PC during the installation. As already mentioned in Sect. 2.2, I've got the permission to deliver a test data set for this book. Thanks to Mr. Nandakumar from the Signal and Image Processing Area, Space Application Centre, ISRO, India and to Mr. Dabrowski from GEO-SYSTEMS Polska for their help to get these permissions!

So, if you like to see an example, just start LISA FFSAT (Start > Programmes > LISA > FFSAT) and select the project TUTOR_6. This programme is very similar to LISA FOTO, therefore we will only take a look to the main differences:

Instead of Pre programmes > Camera definition you may use Pre programmes > Sensor definition. In our case we use data from the Indian Cartosat-1 satellite. The reference system (ground control points) is UTM, ellipsoid WGS 84, zone 34, width 6°. To get a "standard" photogrammetric look and feel in the stereo viewing, the images were turned by 90° clockwise—so this option must be activated here. The sensor data file is already prepared for you and is named SENSOR_1.CMR. Also the control point file and point sketches for each control point are prepared and were stored onto your PC during the installation.

Prepared files: SENSOR_1.CMR, CONTROL.DAT, *.QLK.

As mentioned before, the exterior orientation is defined by the RPCs and must be enhanced by the measurement of control points. The RPC data are contained in the files BANDA_RPC.TXT and BANDF_RPC.TXT. If you like you may carry out the control points measurement using Pre programmes > Orientation measurement. This option is quite similar to the measurement of the exterior orientation in LISA FOTO. The example images used here (already imported from TIFF, converted from 16 to 8 bit radiometric resolution and contrast enhanced) are named 201.IMA and 202.IMA. Finally, Pre programmes > Define model must be run.

Prepared files: 201.IMA, 202.IMA, 201.ABS, 202.ABS, 201202.MOD.

Now the stereo model is ready for use. In particular this means that you can measure 3D co-ordinates and objects in stereo mode using for instance the anaglyph method (option Processing > Stereo measurement, see also Sect. 4.5). Further, you can create a surface model by stereo correlation (see Sect. 4.6) and an ortho image (see Sect. 4.7)—really in the same way as you already know from earlier tutorials. For details please read the relevant chapters in this book or in the programme descriptions (c:\program files (x86)\lisa\text\lisa.pdf).

9.2 Stereo Images from Flatbed Scanners

From high to low, from very large to small objects… do you have a flatbed scanner? Remember that we already talked about such instruments in Chap. 3. There we have used the scanner for what it is constructed: To scan two-dimensional

Fig. 9.2 Geometry of flatbed
scanners

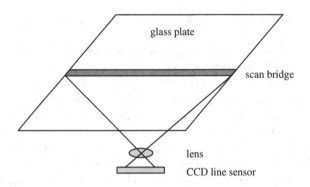

"objects" like aerial photos. In such a case we can simply assume that we have
parallel projection.

In this chapter we want to use a flatbed scanner for digitising small 3D objects.
Of course the depth of focus is not very large, but if we have objects with a
thickness of some millimetres the images will be sharp enough. Before going on
let's look a bit closer at the geometry (Fig. 9.2):

Similar to the cameras operated on satellites here we also have central-
perspective geometry within the scan line but parallel projection from line to line
(along the movement of the scan bridge). In this chapter we don't want to develop a
geometric model for 3D measurement but only show how we can use a scanner to
construct 3D views using the anaglyph method known from Sect. 4.4.

We will use the central perspective within a line in the way that we first put our
object near to the left border of the glass plate/scan bridge, then near to the right
border. Save both images in a common format like JPG. For the remaining steps we
will use LISA BASIC and some tricks:

- Load one of the images into the raster image display. Go to Additionals > Info
  image. In the Geometry section click onto Formal, then OK. Then save the
  image (File > Save or click onto the diskette button). Now the image is
  "geocoded".
- File > Edit project, then Reference raster. Load the image, then OK. Now the
  project area is equal to the image size.
- For both images: Load the image into the image display, then select
  Radiometry > Resolution (bits) > 24 → 3 × 8. Now we have the three base
  colours (red, green, blue) separated for each image.
- Image processing > Rectification > Image to image. Load the red colour
  extracts of the first image (formal geocoded) and the second image (in most
  cases, the red band has a maximum of contrast). Now measure corresponding
  points in both images, then rectify the second image using the option project
  limits. For details to this step please look at the programme description of
  LISA BASIC. If everything is done you have a rectified red colour extract of the
  second image with the same size (No. of lines and columns) as the first image.

- **Image   processing > Matching > Others,** then 2×8 bit → **anaglyph image.** Use the "formal geocoded" first image and the rectified second image. The result is an anaglyph image which you can view stereoscopic using the red-green glasses.

On the Springer server you can find an example (…\tutorial_5\anagl_3.jpg). The object is a handcrafted golden cross from Katakolon, Greece. The original size is about 27×22 mm, the central "rose" has a thickness of about 3 mm. The cross was covered with black drapery to get a dark neutral background, and scanned with 600 dpi.

9.3   A View into the Future

Let's finish our tutorial with some speculative words about the future of the fascinating methods we have seen. Will we still need (traditional) photogrammetry in 10 or 20 years? There are at least two points of view.

The first one: We can recognise that satellite-born *image data* available for civilian use increase in ground resolution and decrease in costs since many years. Nowadays, images with and below 1 m ground resolution are available on the market and also stereo images as we saw just before. *Elevation data* of high precision are collected using laser scan techniques, world-wide data sets are offered free-of-charge, for instance the GTOPO30 data, and will be improved more and more, for instance from data collected during the stereo MOMS and the SRTM missions. From this point of view, it is a question if images, taken with aerial cameras operated on airplanes, will have a future. If they are needed, they will be taken with digital cameras, simultaneously registering the projection centre co-ordinates and the rotation angles (φ, ω, κ) using GPS and IMU techniques. In some years a completely, real-time processing of all data "on the fly" may be the state of the art—after landing, products like ortho images and elevation data are ready-for-use.

The second one: There exist millions of aerial images world-wide. For any kind of historic evaluation these are of an immense value! Any kind of time-series research will need them. Examples:

- Changing of the size and thickness of glaciers, indicating climatic changes
- Destruction of tropical forest in many countries
- Increase of areas used for settlements and roads
- Reconstruction of destroyed historic buildings
- Detection of dangerous points and areas from (historic) images taken after a war: bombs, mines, destroyed tanks and others.

And of course, as we saw, images from digital cameras must be processed in the same way like those from traditional film cameras after scanning. This will be true particularly in close-range photogrammetry also in the future. Concerning the fact that simple photogrammetric work is possible and inexpensive using a standard PC, software and a digital consumer camera, this technique may find a lot of new fields and applications.

Chapter 10
Programme Description

In this chapter we will give brief descriptions of LISA BASIC (only the relevant options useful for photogrammetry), LISA FOTO, BLUH and LISA FFSAT. Of course, many of it you will already know if you have followed our tutorials. Nevertheless it might be good to have a summary for a quick reference. Thanks to Jörg Jozwiak, Berlin, for the translation of this chapter from German to English!

10.1 Some Definitions

- In due course co-ordinate values x and y will always refer to a mathematical, left-hand system, that means "x to the right, y to the top".
- *DTM* generally refers to a raster image of 16 bits depth in the LISA format.
- In digital image processing the expression *image co-ordinates* refers to pixel positions (row/column), while in classical photogrammetry it indicates the co-ordinates transformed to the fiducial mark's nominal values. For differentiation, the expression *pixel co-ordinates* will be used always and only in the context of digital image processing.
- The area being covered by stereo images (image pair) will be called *model area*.

10.2 Basic Functions

As opposed to most digital stereo workstations (DPWS), regarding the direction of the rays, LISA FOTO operates not "top → down" but "bottom → up". Its conception is based on a number of reflections undertaken already several decades ago, for example in connection with the development of correlators (see Houbrough,

© Springer-Verlag Berlin Heidelberg 2016

W. Linder, *Digital Photogrammetry*, DOI 10.1007/978-3-662-50463-5_10

1978, or Konecny 1978, for instance). Those ideas find a digital application in this software.

The orientation of the stereo model in LISA FOTO also differs from customary modes. Instead of the classical division into three parts (interior, relative and absolute orientation) it features an independent orientation for every single image. Therefore it does not comprise a relative orientation in a classical sense—following the interior and exterior orientation of every individual image, only a simple model definition is necessary (see Sects. 4.2, 4.3).

In all programme parts in which you have to measure within a single image or a stereo model, the principle is "fixed measuring mark(s), floating image(s)" like in analytical plotters. The movement of the image(s) can be done with depressed central mouse button, the arrow keys or by moving the marked area in the overview image.

10.3   Limitations

Please note that the software available with this book is a selection of special versions with reduced functionality. Information about unlimited versions can be given on demand.

Images are limited to a maximum size of 20 MB each; this allows for instance the processing of standard grey scale aerial photos with a scan resolution of 300 dpi (about 84 μm, see Sect. 3.2) or images from simple digital consumer cameras.

For aerial triangulation, a maximum of 30 images per block in a maximum of three strips can be handled simultaneously. For image co-ordinate measurement, the number of points are limited to 900 per model and 10,000 in total.

10.4   Operating the Programmes

File names including the path may consist of up to 120 characters.

The button ⌷ opens a file selection window.

Directions are reckoned clockwise from North = 0°, therefore East = 90°, South = 180° etc. Numerical values require a decimal point, not a comma (for instance 3.14 instead of 3,14).

Instead of the button OK offered in each input window the Enter key may be used. Instead of the Cancel or Back button it is possible to use the ESC key.

10.5 Buttons in the Graphics Windows

Display of the image parts, stereo model:

L︱R Side by side left—right
🔲 Overlaid using the anaglyph method (red-cyan)

Size of the display:

🔍 Reduce
🔍 Normal size, 1 image pixel = 1 screen pixel
🔍 Enlarge
🔳 Centre

Form of the measuring mark(s):

▪ Point
+ Cross
✕ Cross diagonal
⊙ Circle with centre point

Colour and size of the measuring marks can be changed and stored.
Measure, register:

  Create sketch
  Distance
◁? Angle
○? Centre and radius of circle
  Polyline
↘. Go to position
r= Correlation coefficient
✓ Ready
✕ Cancel

10.6 LISA BASIC

Starting LISA, a *project* has to be defined. With this, a working directory, an optional image data base, co-ordinate frame (minimum and maximum for x, y and z) and a pixel size will be specified. In the working directory all input files are searched for and all output files are stored by LISA.

The project definition files have the extension. PRJ and are located in the programme directory.

10.6.1   File > Select Project

Corresponds to a new start of the program. Alternatively, the last used project can be taken, one of the existing projects can be selected or a new project can be defined.

10.6.2   File > Define Project

The following parameters have to be defined:

- Name of the project: From this, the definition file (extension .PRJ) will be generated.
- Working directory (folder).
- Image data base (optional).
- Co-ordinate range in x, y and z. The button Reset puts these values to the maximum possible ones. In such a case the limits of x and y are without meaning. A unique value range for z is very important for the generation and matching of DTMs!
- Pixel size (geometric resolution in the object space).
- Length unit (μm, mm, m or km).

> The pixel size and the range of the z-values are valid for all data of a project! Therefore these values should definitely be chosen carefully!

The co-ordinate limits can be overtaken from an already existing project, a geo-coded raster image or a vector file (buttons Take over, Reference raster or Reference vector).

10.6.3   File > Edit Project

The parameters of the actual project with exception of the project name and the working directory can be modified here.

10.6.4   File > Import

Input formats, vector: AutoCad DXF, ASCII files with any sequence (e.g. CSV), dBase DBF, ArcInfo E00, MapInfo MIF/MID. Input formats, raster: Arc/Info ASCII, GTOPO30, SRTM (each 16 bits).

AutoCad DXF: All lines up to the entry ENTITIES are skipped. Co-ordinates which follow the entries LINE or POLYLINE (and then VERTEX for several times, until SEQEND) are rated as points on one line. Co-ordinates after VERTEX without previous POLYLINE or after POINT are rated as single points. All other entries will be ignored.

ASCII any Sequence: Universal import filter for files containing the required entries x, y, z as well as optional the number for each reference point within one line, but in an unusual order or sprinkled in among other pieces of information. Example: In each line the entries

point No. code_1   z   code_2   x   y   operator

are stored in. For the processing in LISA, however, the data must look like this:

point No.   x   y   z

Accordingly, the position of the number has to been set to 1, of the x-value to 5, of the y-value to 6 and the one of the z-value to 3. For instance, CSV files can be processed here. For a maximum of 15 numerical entries within one line and a maximum line length of 200 characters.

Dbase DBF: The input file must contain each a field for the x and the y co-ordinates. From the remaining fields, a numerical one must be chosen from which the z-values are taken. Values for x, y and/or z which cannot be read are set to -999999.

ArcInfo E00: Entries of lines (part ARC) as well as points (parts CNT or LAB) will be adopted. Lines will be given a code according to the so-called coverage-ID (if lower than 5001, increased by 5000). Individual points are given the code 1 and the so-called centroid number from the input file.

MapInfo MIF/MID: Adopted are entries of the types POINT, LINE, PLINE, PLINE MULTIPLE and REGION. A numerical field of the MID file can be used to give the z values, otherwise the z values will be set to -999999.

GTOPO30: Beside of the image file (extension DEM) a file of the same name but with the extension HDR (additional information) must exist. Optional, the part of the DTM situated within the project area can be transformed to one of the projections Gauss-Krueger or UTM.

SRTM: Optional, the part of the DTM situated within the project area can be transformed to one of the projections Gauss-Krueger or UTM.

10.6.5   File > Export

Output formats, vector: AutoCad DXF, CSV, dBase DBF, ArcInfo E00 (only lines), MapInfo MIF/MID, HP-GL 1. Output formats, raster: Arc/Info ASCII 16 bits, DAT/vector.

Note: In the vector data display, vector data can also be converted into the raster image formats BMP or JPG or transferred into other programmes using the clipboard (Additionals > Copy).

CSV: The contents of the input file will be exported without point numbers and codes and with commata as separators.

Dbase DBF: A DBF-file with three numerical fields (x-value, y-value and z-value) is generated, each of which having 12 digits with 3 decimals; the name of the third field may be altered. This file can be used to generate an attribute file, for instance by replacing the z-value field by another and/or adding further fields.

HP-GL: For output on a plotter. The file can be viewed in the graphics display of BLUH.

DAT/Vector: The input image must be geo-coded. Output is a vector file with the z-values derived from the input image.

- Single point or profile data: The x/y-values are taken from an input vector file. For profiles, the interval (=distances between points) must be defined.
- Grid data: Defined by the grid width. As an option for each pixel of the input image a point may be issued (All Points).

Points within the raster image for which no information is available (colour value = 0) will not be transformed.

10.6.6   Vector Data > Control Points

For creating or editing a control points file. The number of control points needed depends on the transformation method; a maximum of 900 points can be processed there. The co-ordinates have to be given in a Cartesian system (e.g. Gauss-Krueger or UTM) and should be well distributed within the area. If the map gives only geographic co-ordinates (longitude, latitude) these values must first be transformed to Gauss-Krueger or UTM (see Vector data > Projections).

10.6.7   Vector Data > Define Symbols

Single points of the codes 3501 ... 3600 can be connected with vector symbols. There are 10 standard symbols (circle, cross etc.) which can be modified in size. A file named VEC_SYMB.DAT will be created in the actual working directory.

10.6.8 Vector Data > Projections

This option is for the transformation of vector data between several co-ordinate systems. For example, non-cartesian geographic co-ordinates (longitude, latitude) can be transformed to Cartesian (metric) systems like Gauss-Krueger or UTM. Projections:

- Geographic → Gauss-Krueger
- Gauss-Krueger → Geographic
- Geographic → UTM
- UTM → Geographic

Geographical co-ordinates have to be entered in the order x = longitude, y = latitude and in the unit decimal degree or degree/minute/second as one number (example: 7° 12 min 24 s is entered as 71224). Further: Eastern longitude → positive values, western longitude → negative values. Northern hemisphere → positive latitudes, southern hemisphere → negative latitudes or option **Southern hemisphere** activated. For the entering of the zone: This may be given explicitly in case for example co-ordinates of another than the actual zone in question shall be calculated. Otherwise the zone is calculated automatically from the longitude value.

Gauss-Krueger and UTM data have to be entered in the order x = Easting, y = Northing resp. x = East, y = North and in the unit meter. For UTM → Geographic the zone must be defined. The Northing values of UTM co-ordinates of the southern hemisphere are according to definition to be entered with an addition value of 10,000 km and option **Southern hemisphere** activated or as negative numbers. If an area is located both north and south of the equator, either all north values should be increased by 10,000 km and the option **Southern hemisphere** activated or values south of the equator must be negative (example: Ecuador).

For special applications, the position of the co-ordinate centre in degrees as well as the corresponding co-ordinates in metres can be defined explicitly.

10.6.9 Vector Data > Vector → Raster

Situation overview: Points can be entered with their numbers or height values.

Area filling: Precondition besides a vector file containing the geometry in the shape of one or more polygons (closed polylines) is an attribute file (DBF) with anchor points (one point per polygon). Choose a numerical field out of these attribute files. The value range can be joined to classes (intervals) using one of the following methods:

- Equal distances with given number of classes
- Equal distances with given width of classes
- Equally distributed, number of classes pre-defined

- According to natural breaks. The size of the interval which shall be interpreted as a break can be defined using the parameter class width
- User defined by a limit values file, containing in each row the data "from", "to" and colour value.

Example for a file with limit values:

```
12.8   20.7   6
20.7   28.4   7
```

The values between 12.8 and 20.7 will be given the value 6, those between 20.7 and 28.4 the value 7. The parameter "to" is inclusive, so in the example above a value of 20.7 would belong to the first class (6). Then the programme fills the polygons with the colour value defined by the class number.

In the special case that all z values are integers in the range between 0 and 255, instead of a classification the option **Class = z value** can be used).

Note: It is important that the single areas are bordered by closed polylines. Especially these have not only to be closed optically but mathematically as well (the co-ordinates from the first and the last point are identical)! In order to meet this precondition, for example you can use the vector graphics display and then **Edit > Move,** or when digitising, the button **Close.** For the area filling, colour values between 2 and 254 are used.

Free cut areas: If the input file consists of one or more closed polygons (code 9008) and of one or more starting points for deletion (code 4007), only the marked areas for not free cutting of the input image will be taken over into the output image (masking function).

Note: The vector graphics display offers more possibilities of editing (see there).

10.6.10   Image Processing > Two-Dimensional Histogram

This gives an idea about the degree of correlation of two images of the same size (8 bits each). The brightness of the grey values corresponds with the frequency. The highest frequencies can usually be found around the main diagonal. The more culminated and precise they are the stronger is the correlation between the two channels. A further indication gives the correlation coefficient shown in the lower part of the histogram.

10.6.11   Image Processing > Rectification

For the orientation or rectification of an image an equation system is set up which enables an assignment between object- and pixel co-ordinates. Concrete: With the help of the control points' co-ordinates and the corresponding pixel co-ordinates, the coefficients a_i and b_i of the equation systems

$$x' = a_0+a_1x+a_2y \qquad\qquad\qquad y' = b_0+b_1x+b_2y$$
$$\text{(affine transformation)}$$

resp.

$$x' = a_0+a_1x+a_2y+a_3xy+a_4x^2+a_5y^2 \qquad y' = b_0+b_1x+b_2y+b_3xy+b_4x^2+b_5y^2$$
$$\text{(2nd order polynomials)}$$

are determined. An adjustment using the least squares method will be applied in case of over-determination. In the first case at least 3 points are required, in the second case at least 6 points. If only 2 control points exist, a similarity transformation is calculated.

In case of over-determination, the residuals (remaining errors) in each point as well as their standard deviation are shown. If a single value shows a strong deviation from the arithmetical mean with sufficient over-determination the transformation should be done without this point. For this use the option (De)activate. The number of this point now is set to the negative value. With a second time marking the point it can be re-activated.

Further methods (without adjustment): Projective transformation (4 points), local/rubber sheet (4 ... 900 points). The latter one is suitable for instance for images with non-regular distortions (like historical maps etc.).

For the calculation of the rectified image there exist different methods for the determination of the colour values ("resampling"), LISA offers the two most important:

Nearest neighbour: Fast, keeps the original colour values, but gives a coarse appearance especially for enlargements.

Bilinear: A bit slower, smoothes the colour values similar to a mean filter by taken into account a 4 pixel neighbourhood. The result is "nicer" than before but contains additional colour values (mixed pixels).

So, in all cases where the original colour values are of interest for further operations, the first method will be the better. On the other hand, if the optical impression is of higher priority, choose the bilinear resampling.

Single image

Input: Image and control points file. Method: Move the image with the middle mouse key (or the mouse wheel) pressed until the control point selected in the list below is in position with the measurement mark, then press the left mouse key to save the point data. After a click onto the Ready button the rectification will start (see below, Numerical). The orientation parameters are stored in a file named <image name>.ORI.

Image to image

If a geocoded image exists, another image can be rectified to the geocoded one by a direct measurement of reference points. Middle mouse key: Simultaneous movement of both images. Right mouse key: Movement of only the right image. Left mouse key: Registration of points. Method: Move the images with the middle mouse key (or the mouse wheel) pressed until the point of interest is in position with the measurement mark in the left image. Now move the right image with the right mouse key pressed until the respective point is in position with the measurement mark in the right image, then press the left mouse key to save the point data. After a click onto the Ready button the rectification will start (see below, Numerical).

Corner co-ordinates

After input, the image will be rectified (see below, Numerical).

Numerical

For the rectification of an image, a special control point's file is used containing the values [Point No., x, y, row, column] of each point. Such a file named GEOCOD_R.DAT will be created from all of the methods mentioned above. Example:

```
        1 32489700.000   6071025.000   702.000        969.000
        2 32498100.000   6052275.000   829.000       1171.000
        3 32504700.000   6037275.000   902.000       1328.000
        4 32528100.000   6072225.000  1143.000        990.000
    . . .
32422730.209 32544204.406              co-ordinate frame in x
 6000748.517  6163394.639              co-ordinate frame in y
```

The co-ordinate frame for the output image is to be set—the option Project limits will use the values from the project definition.

Note: All transformation methods described above lead in a mathematical view to a transformation of one plane to a second plane parallel to it. For the rectification of aerial images better use the option Ortho image in LISA FOTO.

Adapt

Only for geocoded images and DTMs. The selected image will be adapted to the geometry of the actual project (see File > Project definition): The image content will be limited to the co-ordinate frame of the project, the geometric resolution (pixel size) will be re-calculated to those of the project, and in case of a DTM the height resolution will be adapted.

10.6.12   *Image Processing > Mosaic*

For up to 5 rectified (geo-coded) images or DTMs. In overlapping areas the colour values can be overwritten or averaged.

It is possible to use in the first input field (Image 1) "wild cards". For example, if Image 1 = TEST*.IMA, then the files TEST1.IMA, TEST2.IMA etc. are used for the mosaic. This option is limited to 100 images.

10.6.13   *Image Processing > Classification*

Outside of photogrammetry—see programme description LISA.PDF if need be.

10.6.14   *Image Processing > Matching*

In contrary to the combination of images as mosaic (see above), with matching the images are overlaid and the colour values of the output image calculated from the corresponding pixels. For this, the images must have the same dimension (No. of rows and columns).

The chosen method is done pixel for pixel, for example addition: Colour value of the output image = colour value image 1 + colour value image 2.

Note 1: The entering of a third image is only necessary for some options (see below).

Note 2: For more sophisticated matching (e.g. with more than two input images or using free-definable formulas) see the option Analysis > Formula calculation.

The different possibilities in detail:

Addition

- Clipping at white: Sum colour values are limited to the value 255 (white).
- Scaled sum: The resulting colour values are transformed linearly onto the range 0 ... 255. Options: Either the actual value range between the minimum and the maximum or the theoretical possible value range (0 ... 510) will be transformed. The latter case corresponds to the arithmetic mean.
- Weighted addition: "Double exposure". The weight of image 1 can be set between 1 and 99 %, image 2 will be weighted with the difference of this value to 100 %. See also Others > 2 × 8 bit → 24 bit.
- To gap (image 1 = 0): Only where no information is given in the first image (colour value = 0) the colour values will be taken over from the second image, either those from the first image will be used.

Subtraction, masks

Remark: A binary image is called a mask. It may be used to cut specific content off other images (see the last two options below).

- Clipping at black: Different colour values are limited to the value 0 (black).
- Scaled difference: The resulting colour values are transformed linearly onto the range 0 ... 255. Options are the same as in the addition.
- Absolute sum from image 1 to image 2.
- Image 2 as mask, remains maintained: Wherever there is a colour value higher than 0 in image 2, the point will be left blank in the output image (set to 0). Otherwise the colour value from image 1 will be taken over.
- Image 2 as mask, remains away: Exactly the other way round.

Division, ratio

- Arcus tangens of image 1/image 2, increased to mean colour value.
- Scaled quotient: The resulting colour values are transformed linear onto the range 0 ... 255. The quotient can be limited to a maximum value.
- Vegetation (NDVI): e.g. for LandSat-TM bands 3 and 4; as formula: (image 2 − image 1)/(image 2 + image 1) with image 1 represents the visible red and image 2 represents the near infrared.
- Snow/NDSI: Like before, but image 1 visible green, image 2 middle IR.
- Moisture/NDMI: Like before, but image 1 near IR, image 2 middle IR.

Others

- Minimum, maximum: The minimum or the maximum value from the colour values of the two input images is used pixel-wise as output colour value. Note: The arithmetic mean can be derived with the addition (see above, scaled sum).
- Directed cosine: Is calculated as quotient of the colour values of a channel to the radiometric distance of the respective point from zero (colour value = 0). The individual channels and the selected channel are to be entered.
- Colour composite: From three channels (raster images of 8 bit each) a 24-bits colour image is generated. To aim for a good optical impression it may be useful to bring the single input channels to their maximum contrast first (in the image display, there Radiometry > Histogram > stretch).
- True colour image LandSat TM: Like before, but designed particularly for the channels 1, 2 and 3 of image data originating from LandSat-TM satellites. A particular adaptation of the histograms generally leads to a good optical impression.

- 2 × 8 Bit → 24 Bit: Similar to the "Double Exposure" (see above) but designed particularly for 8 bit colour images (for instance, coloured height representation plus shaded relief, DTM). The result will be stored as a 24 bit image.
- 2 × 8 Bit → Anaglyph image: Like before, but a "3D image" will be created which may be viewed using red-green glasses. Prerequisite are two images from the same object but taken from slightly different positions (one a bit more from the left, the other a bit more from the right).

10.6.15   Image Processing > Area Symbols

For the creation of area patterns. The symbol size (number of rows and columns) and the desired colour value are to be entered; moreover, a symbol already existing can be loaded as a draft. An orientation raster appears left on the screen, a test image right. Depending on the selected tool, points can be set or deleted with the left mouse button. You should take into account that with the continued rowing of the symbols a sensible pattern should result—this can be checked in the test image. The finished symbol will be stored with a number between 101 and 200, the extension is SIG.

In the image display, symbols can be used via **Palette > Area pattern** (see there).

Besides the area filling it is possible to assign the symbols to single points holding the respective code within the vector raster conversion (e.g. code 3501: the symbol file 101.SIG is searched for and if needed be entered into the raster image).

Note 1: An image in which the colour values were exchanged by an area dithering is highly suitable for a mathematical matching with a same sized grey- or colour image in the sense of a raster overlay. Then the grey- or colour tones are visible through the area dithering (cf. above, **Matching > Addition > Clipping at white** or **Matching > Subtraction > Clipping at black**).

Note 2: See also the option **Radiometry > Filter2/binary > Floyd-Steinberg** in the image display as an interesting alternative.

Note 3: Self-created symbols (e.g. 101.SIG) are not stored within the working directory but within the common directory c:\users\public\lisa\sig.

10.6.16   Terrain Models

Almost every type of 3-dimensional co-ordinates can be processed. Usually the aim is to produce and process a *height* model. In order to realise this, a network of well distributed points (=as densely and homogeneously as possible) with the x, y and z (height) co-ordinates being known, is required over the working area.

Note that the quality of a terrain model, which means the extent of an agreement with the real terrain, depends primarily on the density and distribution of the input data (reference points). It is therefore absolutely necessary to take the terrain topography into account while collecting the data. A general rule is: The more undulated the terrain the denser the point network should be. Terrain elements such as edges of steep faces, ridges or V-shaped valley floors should be measured separately and defined as break lines. Local minima and maxima (single points which appear to be higher or lower than their surroundings) have to be entered in the reference point file, too.

The data must be available as vector file; the number of reference points is not limited. The parameter code is used to define the following meanings (see the appendix, vector data).

Example of a border polygon:

```
9008   -999999.     -999999.         1.
   1      1000.        1000.        10.
   2      1200.        1200.        10.
   3      1050.        1600.        10.
   4      1000.        1000.        10.
 -99       -99.         -99.       -99.
... and a starting point for deletion:
4007   -999999.     -999999.         1.
   1      1900.        1900.        10.
```

... this one is located outside the polygon, thus it will be deleted outside.

Important: Border polygons must either be closed, or start and ending point must each be situated outside of the DTM area! Otherwise, the deletion algorithm will "go through a gap" and will delete also other areas. In an extremely case, the whole DTM can be deleted. This is the fact if at the end of the DTM calculation the error message "All grey values equal" (=zero) appears.

10.6.17   Terrain Models > Interpolation

Calculates a digital terrain model (DTM) from the input points as a raster image, scaled to the range 1 … 32767 ("16" bits, actually for technical reasons 15 bits).

Moving average: Universal, quick method. Best results when reference points are fairly well distributed and densely located. Very suitable for the quick visualisation of an overview and the realisation of major mistakes in the reference point data. Especially close to linear elements (break lines) this method tends to a formation of plateaus.

Moving surface: Universal method with mostly smooth contour course in flat terrains. Mostly a high accuracy, lower tendency towards a formation of plateaus but also a bit slower. Automatic change between polynomials of 2nd order, tilted plane and horizontal plane. Instead of the automatic switching the method can be fixed to tilted plane for special occasions.

Trend surface: If the number of input points lies between 5 and 900, a trend surface can be calculated through these points. This is a 2nd order polynomial or optional a tilted plane.

Parameters:

The maximum distance from which reference points are to be searched for starting at a new point (search window size) must be entered (in metres).

For the first two methods, the number of reference points which will be searched in each segment for interpolation can be defined between 1 and 10.

A smoothing of the DTM surface may be reached with a mean or Gauss filter of variable window size (standard = 5×5), which is useful, for example, for the generalisation of contour lines. Significant single points (code 4006) as well as "hard" break lines (code 9010) are not touched by this.

Border polygons are closed automatically. In special cases where a border polygon contains several parts and no unified circulation do exist ("spaghettis"), the automatic closing can be turned off.

10.6.18   Terrain Models > Filtering

Fill local minima: Within a selectable window size all pixels which are lower than the border pixels of the window will be set to the minimum border pixel value (filling depressions without runoff).

Remove peaks: The DTM will be subdivided into tiles of the defined size. In each tile, only the pixel with the lowest grey value will be maintained. From these points, a new DTM will be created by an interpolation with tilted planes. All points of the original DTM which are higher than respective high of the new DTM plus the given threshold, will be eliminated (for instance trees, buildings), the caps then are filled by interpolation.

Filter: The DTM can be filtered with a mean, Gaussian or self-defined filter with variable window size.

10.6.19   Terrain Models > Contour Lines Vector

The contours are calculated as a vector file (No., x, y, z). The smoothing of the lines is done via subpixel interpolation, the effect of this can be handled in the data reduction running directly after the line calculation: The lower the tolerance value, the lower is the smoothing and the higher the number of output points (see also the vector graphics display, there Edit > Thinning).

10.6.20   *Terrain Models > Numerical Evaluation*

In general: All calculations refer to the actual DTM-area. Free cut areas (grey value = 0) from the interpolation are not taken into account in the computation.

Total area: Calculates ground and real surface area of the entire DTM. "Ground surface" means the parallel projection onto the x-y-plane and "real surface" the actual three-dimensional surface.

Intersecting surface: Calculates the total area of all regions situated on or above a given height.

Volume: Calculates the volume between the surface of the terrain model and a reference plain of constant height. The output is done separately for the volume above and below the reference plain. The latter one corresponds to the filling volume.

Volume differences: Calculated from a difference-DTM. The output is done separately for volume increase and decrease.

Statistics: Relative [%] and absolute areas for heights, inclination and aspects. For heights and inclinations interval sizes are to be given.

Take into consideration the resolution in all computations! The values of x, y, and z in the reference point file from which the DTM was interpolated have to be in the same units. If the values of x and y are in km, for example, and those of z in m, then the result of the volume computation is definitely wrong!

Note: The option Volume is only suitable for "true" DTMs, the option Volume differences only for difference-DTMs. For the each opposite case the results will be wrong!

10.6.21   *Terrain Models > Matching*

Addition

Add two DTMs of equal position and size.

Differential DTM

Important: The difference-DTM will only deliver expressive results, e.g. for the calculation of volume changes, if the following conditions are met:

- The pixel sizes, height ranges as well as the co-ordinates of the corners are the same.
- If free cut areas exist, their number and locations must correspond.

Masking

According to a binary image (mask) of the dimensions like the DTM, all pixel positions with a grey value >0 will be maintained in the DTM, all others set to 0.

10.6.22 Analysis

Outside of photogrammetry—see programme description LISA.PDF if need be.

10.6.23 Data Base

Outside of photogrammetry—see programme description LISA.PDF if need be.

10.6.24 Display Raster Image

Input: 8- or 24-bits raster image (format IMA, BMP, JPG, PNG or TIF). 16-bits DTMs are converted to 8 bit internally. Co-ordinates display bottom left in the status line: If the image is geocoded, the values of x and y are displayed, either column and row. At the third position comes the colour value and, if it has a numerical meaning (for instance the terrain height), at the fourth position the respective z value. If 24-bit images are used the colour intensities (red, green and blue) are shown in the 3rd, 4th and 5th position. If a text reference with the same name as the input image exists, instead of the z value the text belonging to the respective colour value will be displayed (e.g. land use).

Please note: The image display in LISA BASIC offers a lot of options as described below. In LISA FOTO and LISA FFSAT only few of them are available.

File

Open, Save, Save as, Print.

Palette

Normal, Negative, Colour 1, Colour 2, Open, Individual, Area pattern, Brightness/Contrast, Flood.

Open: To load an existing palette.

Individual: The colour of each pixel value (0 ... 255, click onto the square) can be adjusted individually, mixing the primary colours red, green and blue (additive mixture). The option Continuous (symbol ◢) calculates all values between the colours to be provided (from ... to). The option System starts the Windows colour mixing option for the actual colour value.

Area pattern: Colour values can be replaced by area patterns. The colour value range and the distance between the lines or points in pixels are to be entered. Instead of area symbols, raster images (e.g. scanned graphics) can be used.

Flood: Especially useful for DTMs—the area below the defined grey value will be displayed in a blue colour.

The original palette can be restored using the button Reset.

View

Reduce/Enlarge: With mouse wheel or by setting the percentage value.

Move: With pressed middle mouse key. Further, the image detail displayed in the overview image (see below, **Additionals**) can be moved using the left mouse key, then moving the image simultaneously.

Rotate by 90, 180 or 270°, flip left-right or up-down with the right-hand buttons.

Optimal: Maximum zoom factor in order to display the entire image.

Window: Defines a part of the image by a window drawn with the mouse.

New drawing or button Reset: Like when starting the image display.

Profile: The profile trace is determined by 2 ... 100 points of a vector file which must be defined. Creates a colour value profile over the input image or a height profile (if the image is a DTM). In case of a DTM, optional in each point a lateral profile can be created.

3D-view: Available if the image is geocoded and a DTM with the same ground resolution exists. You can select azimuth, inclination and exaggeration. The result can be stored as new image. As an alternative, from the DTM a wire-frame representation can be calculated.

DTM

If the input image is a 16-bit DTM, additional views can be created:

Height steps: The elevations will be grouped according to the desired equidistance (interval in z-direction) with the possibility to limit the range in the output image. The result represents equidensites of first order.

Contour lines: The desired equidistance must be entered. Count lines can be created (for instance each fifth) and will be drawn in red. The output can be limited to a particular height range. The lines can be coded in colour values. If only a part of the contour lines is to be shown, the area below the lowest contour line can be hatched ("flooded").

Shading: An illumination with parallel incident rays is simulated. The azimuth (north = 0°, clockwise up to 360°, continuously) as well as the tilt angle (10° = flat up to 80° = steep) are to be defined.

Aspects: The aspects N, NE, E, SE, S, SW, W, NW and "not tilted" are represented in different colours.

Visibility: You need a vector file with target points which must be located within the DTM area and above the terrain. For instance, this might be radio antenna positions, and the question is if there are areas which will not be reached, caused by the terrain. A radius must be defined (e.g. the maximum distance of the emitter). The programme creates an image showing all areas with free sight to the targets in a dithered representation, colour coded in respect to the distance.

Measure

Registration: For registering of co-ordinates in the terrain system. For a maximum of 20,00,000 points.

The input of a z-value is interesting when digitising contour lines for example. If the z-value is unknown or irrelevant 0 might be entered instead.

A code between 1 and 5000 represents single points, 5001 ... 9999 polylines. Codes between 3501 and 3600 can be linked with single point symbols (see Vector data > Define symbols for instance).

The registration of data is carried out point after point via pressing the left mouse key. With the right mouse key the co-ordinates of an already digitised point (e.g. on a polyline) is to be caught. Further options offer the buttons Close, Interrupt line and Delete.

Note: If you would like to collect data for digital terrain models, please note that the codes are relevant. Contour lines should be digitised with the code 9009 (soft break line).

Which kind of co-ordinates will be registered, results from the kind of loaded image:

- The image is geo-coded: Object co-ordinates will be given as an output.
- The image is not geo-coded, no orientation was done: Column and row positions (pixel co-ordinates) are given as an output.
- The image is not geo-coded but an orientation took place before: Object co-ordinates calculated by a transformation will be given as an output.

Area/Perimeter: Areas of any shape may be surrounded (measurement of points with left mouse key); after pressing the right mouse key the area will be displayed. For each area a minimum of 3 points has to be measured. The result is issued with suitable units (m^2, acres, km^2, resp. m or km). Results of additional terrain measurements may be added or subtracted.

Slopes/distances: Two points are to be digitised and their respective height is to be entered. From the measured co-ordinates and the height values the programme calculates the slope inclination in degrees and % as well as the horizontal and spatial distance between the two points in the terrain unit (metres).

Polyline: All points of a polyline are digitised one after another; the last one is clicked on with the right mouse key. The total length in the terrain unit (metres) will be shown.

Analysis of the grey values within a polygon (see option Classification; 8 bits).

Overlay

Vector graphics: The superimposition's colour value may be selected. Individual points may be issued alternatively with the corresponding number or height and with a dot mark (small square).

Attribute data, Photos/Texts: Applies for geo-coded images only! The positions for which information is available are indicated by small squares. Clicking onto such a square induces the display of the relevant data set, image or movie. Before starting further operations or closing the window, the right mouse button is to be pressed once.

Radiometry

Stretch histogram: Improves the contrast by linear stretching.

Histogram equalisation: Improves the contrast by creating a standard distribution of colour values.

Filter 1 (Colours)

Mean (blur, a low pass filter): Forming of the arithmetical mean. The filter works by smoothing and gives a less sharp image comparing with the input.

Edge preserving smoothing: Has the same effect as a mean filter, provided that the contrast (difference between the maximum and minimum colour value in the window) does not exceed the chosen Threshold value. Therefore the output image appears less blurred compared with a simple mean filter.

Median: For the elimination of disturbed pixels (peaks). Assign the mean of the colour values of the neighbours, which are arranged in a rising order, to the central element.

Modal (Majority): If the colour value with the maximum frequency reaches the indicated least frequency within the neighbours, the central pixel will also get this value. Suitable for the optical improvement of classification results.

Gauss: Low pass filter, similar to the mean filter, but the colour values within the window are weighted using the Gaussian density function (normal distribution).

Edge enhancement: The effect of this filter can be set using the parameter Sharpness (0.1 ... 0.9).

Local contrast: Within the selected window size the contrast is enhanced. The effect can be increased or decreased using the parameter Sharpness (0.1 ... 0.9).

Self-defined: The values of the filter matrix, located in a text file with the extension FLT, are to be entered. This for instance consists of 9 values (3 in each row) containing the weight for a 3×3-window (real values also possible). Example:

```
-1   -1   -1
-1   16   -1            (a high-pass filter)
-1   -1   -1
```

The 8 neighbours are each weighed with -1, the central element with 16, the sum is divided by the average value (here: 8) (nomination). If there is a "zero sum filter" (sum of weights = 0) a lifting with the average colour value 127 takes place instead of a nomination.

Negative image: Inversion of colour value range 0 ... 255 to 255 ... 0.

Filter 2 (Binary)

Second order equidensities (edges): Calculates the colour value difference between the current pixel and the pixels at the bottom and to the right. If this value lies below the chosen threshold value the new colour value will be 1, otherwise 0. Thus, depending on the image contrast and threshold value, edges can be detected.

Threshold binarisation: Colour values above the determined threshold value will be set to 1, those below to 0.

Floyd-Steinberg: The image is dithered.

Erosion, Dilatation: Starting from a binary image the objects contained in it (points or lines) are modi-fied in such a manner that their border is either reduced (erosion) or broadened (dilatation) by one pixel.

Open: Erosion followed by dilatation.
Close: Dilatation followed by erosion.

Negative of binary image: Exchanges the values 0 and 1.

Filter 3 (Gradients)

First derivation (gradient): Forms the difference between the colour values of the current pixel and of one of its 8-neighbours, which is determined by the chosen "aspect". The result is a pseudo-relief.

Laplace, blurred mask: High pass filter with strong emphasis on the central element (Laplace: four times with regard to 4-neighbours, blurred mask: eight times with regard to 8-neighbours).

Sobel in x (columns) or y (rows): Linear structures in row or column direction will be worked out with the help of difference forming of the current row (column) to the neighbouring row (column).

Monotony: For each of the 8-neighbours it is ascertained whether the difference between its own colour value and the colour value of the central pixel does not exceed the chosen threshold. The number of these neighbours determines the new colour value, which gives information about contrast resp. homogeneity in the 3 × 3 window.

Variance: The difference between the highest and the lowest colour value in the 3 × 3 window forms the new colour value.

Dynamic: The number of opposite neighbours which exceed the chosen threshold defines the new colour value.

Canny Edge Detector: With this method image edges (lines along strong colour value changes) can be detected.

Steps/Parts

Steps (1st order equidensities): Joins the colour values to groups. The step interval in colour values is to be entered, e.g. 10: The colour values from 0 ... 9, 10 ... 19 etc. are joined together.

Range (from ... to): The colour values situated outside of the interval are set to the value 0. In case of the colour values have a numerical meaning, the interval can also be defined via the corresponding z values. For example 50 ... 150: All colour values under 50 or over 150 are set to the value 0. If on the other hand only the values below 20 and above 200 shall remain, and those between 21 and 199 shall be set to zero, define the parameters "from" = 200 and "to" = 20.

Factor, summand: The colour values of the new image are calculated from the colour values of the old one according to the formula CV_new = factor × CV_old + summand. Values outside are clipped (=set to the extreme values 0 or 255).

Others

Resolution: 24 → 8 bits colour (Floyd-Steinberg), 24 → 8 bits grey, colour layers 3 × 8 bits, 8 → 16 bits ("DTM"), 8 → 24 bits. Colour layers: The file names are as the input file but are expanded by _R, _G and _B.

Noise: This option creates a random noise which will be added to the input image. The amplitude of the noise can be set between 1 and 255.

Fade out: The image borders (width in pixels to be set) will be faded out stepless to the selected colour value.

Geometry

Rotate image by 90, 180 or 270°, **flip image** left-right or up-below.

Image cutout: In a geo-coded image by entering the co-ordinates of the lower left and upper right corner of the segment, in a not geo-coded image by entering the first/last row and first/last column. The pixel size remains unchanged.

Change image size: It is optional whether you enter a percent value or the desired size of the image in pixel. The output image will then be extended row- and column-wise in contrast to the input image. Take into consideration that the image size is changed by the square of the factor. Resampling method: **Nearest Neighbour** or **bilinear** (see **Image processing > Rectification**).

Additionals

Display of overview image, histogram and legend (for 8-bit images or DTMs). The overview image features the position of the detail currently displayed.

Grid: For geo-coded images. Grid crosses or -lines can be drawn into the image in selectable distance and can be labelled with their co-ordinate values at the image borders.

Copy: Stores the image within the clipboard for use in other graphics programmes.

Info image: An existing geo-codification can be deleted using the button **Reset**. The button **Formal** creates a "formal" geo-codification.

10.6.25 Display Vector Graphics

This option can be used for vector files up to 20,00,000 points. You can choose whether you want the points to be displayed with their numbers and/or with their z values. The points and lines may be coloured according to their z values. In the status bar bottom left the object co-ordinates corresponding to the mouse cursor position are displayed.

File

Open, Save, Save as, Print. When exporting into the raster image formats BMP or JPG, the actual displayed part of the file will be stored.

Palette

Normal, Negative, Colour 1, Colour 2, Open. Only available if the option z → colour is active.

View

Reduce/enlarge by setting the percent value. Moving with pressed middle mouse key or pressed mouse wheel.

Window: Define the desired part of the graphics by a window drawn with the left mouse button.

New drawing or button Reset: Display as at start.

3D-view: You can select azimuth, inclination and exaggeration.

Edit

Parts, calculation: The data outside the limits can be taken over into the output file unchanged. A logical AND-connection follows (if x-value in the given range AND y-value in the given range ...) as well as a conversion of the form output value = input value x factor + summand.

Co-ordinates: After clicking on a point with the mouse cursor, depending on the kind of point different parameters can be edited:

- Single point: Number, x-, y-, z-value and code.
- Point on a line: It is either possible to change x, y, z for this single point (option Point) or the code for the entire line (option Line). Finish the edit mode with the right mouse key.

Note: After clicking the OK button, normally the input window will be closed. If it appears a second time, the reason is that there exists another point at the same location.

Points → Lines: Define code (range 5001–9999), then click on point after point with the left mouse key, mark last point with right mouse key.

Move: Shift point with the left mouse key to the desired position, then let go of the mouse key. The co-ordinates of the target point will be taken over exactly— ideal for a secure closing of polylines, e.g. for a following area filling. Finish the shifting modus with the right mouse-key. Note: If after shifting of a point the point seems to remain at its old position, this means that there were two or more points.

Delete: You can choose whether you want to delete single points or entire lines. Click on point(s) with the left mouse key and determine the deletion modus with the right mouse key. If you click on a point (or a line) for the second time it will be restored. Moreover polylines can be separated by a marked point. It is possible to restore this with a repeating click on the point. Further, with the option Window an area can be marked; all points within will be deleted.

Points for deletion are marked with a colour and are not taken over into the output file while storing.

Thinning: For the purpose of thinning polylines, for instance digitised contour lines, using the tunnelling method. A tolerance value has to be provided. Each two neighbouring points determine a straight line. All successive points which fall short of the defined tolerance value won't be taken over into the output file.

Register: After entering z-value and code you can continuously digitise with the left mouse-key. The values for x and y from the mouse co-ordinates are transformed into the terrain system. Close polyline or finish measurement with a click onto the corresponding button. The registered data can be added to the input file or stored in a separate file.

Additionals

Grid: Grid crosses or lines in selectable distance can be added to the graphics and labelled at the image borders with their co-ordinate values.

Copy: Stores the graphics within the clipboard for use in other graphics programmes.

Info Graphics: The number of points in the input file are shown as well as the co-ordinate ranges in x, y and z.

10.6.26   Display Text

This is used for the creation, display, processing and printing of an ASCII file (text file, vector data). The text display is started automatically in LISA at some places, for example after calculating statistical data of a DTM.

10.6.27   Display Attributes

For display, editing or printing of a DBF file. This has to contain in field 1 the x- and in field 2 the y-value (reference- or anchor-point) for each data set and can for example be generated out of a vector file by the option File > Export > Dbase DBF. Limits: A maximum of 50,000 data sets with a maximum of 25 fields.

Rename field: Only the name is changed, the data structure (type, length) remains.

Add field: You have to enter the field parameters (name, type, length). The field will be appended after the last available one.

Delete field: All data are removed.

Data from vector file: The z-value of a vector file that is to be given will be entered into the chosen field of a DBF file according to their x-y-values. Optionally an assignment between (integer) z-values and assigned texts of a text reference can be chosen for the entry of non-numerical data. For this choose a text field, a vector file of the form No., x, y, z and a text reference of the form z, text.

Data from raster image: Based on the x-y-values of the DBF file the corresponding z-values will be drawn from a geo-coded raster image and entered into a field to be selected.

Areas: Required are:

- A raster image (8 bits), containing border polygons of the areas to be calculated with the colour value 1 or 255.
- An attribute file (DBF) which contains for each of these areas an anchor point (x value, y value as first and second field) and a numerical field of a length of 12 digits containing 3 decimals, in which the results shall be written.

The sizes of the areas are calculated and entered into the file.

Select data: Data sets which contain values in two specific fields and are located within certain defined intervals and further on are suitable for a combination will be separated in an output-file. Example: In case the values located in field 3 are between 100 and 200 AND/OR the values located in field 4 are between 35 and 70 the data set will be taken over. Otherwise nothing will happen. If you want to work only with one criteria you can choose the combination AND and set the interval limits for the second field to maximum values (e.g. −999999. ... 999999.). Possibilities to combine: AND (as well as), OR (exclusively, either or), AND/OR, AND NOT.

10.7 LISA FOTO

10.7.1 File > Select, Define or Edit Project

See respective options in LISA BASIC.

10.7.2 File > Import

The input images must be 8- or 24-bit files in one of the formats BMP, JPG, PNG or TIF and should then be imported to the IMA format using this option. Consequently all images of the working directory or all selected ones (press the Ctrl key in the file manager) will be converted to the LISA internal IMA format (batch mode). The output file names are numerical; for instance, the image TEST137A.JPG will be converted to 137.IMA. Options:

- **Rotate** by 90, 180 or 270°.
- If the images coming from different cameras, they can be connected to the respective camera. The image numbers then have 6 digits, e.g. 100137.IMA (camera 1) or 200137.IMA (camera 2). This option is only used for image sequences.

Remarks:

It is possible to work in LISA directly with the formats BMP, JPG, PNG or TIF. Nevertheless, we recommend converting the images into the IMA format with the option described here.

For the photogrammetric evaluation, the image files must have numerical names (maximum 6 digits)! Usually, these correspond with the respective image number. Names like IMG_1022.JPG are not allowed, the file should be renamed to 1022. JPG. Within the import this is done automatically, see above.

Before going on with processing, the calibration parameters should be determined if possible. See the option **Pre programmes > Orientation measurement,** then **Measure > Calibration pattern.**

> Please note: The file extensions for LISA are JPG (not JPEG) and TIF (not TIFF).

10.7.3   Pre Programmes > Camera Definition

Analogue

Preliminary remark: The option discussed here is to be applied in connection with conventional (aerial) photo cameras, providing the original photos were digitised by scanning. After the camera definition for each image an interior orientation has to be carried out (see below). In case the source is provided by a digital camera, the option following next is the relevant one; no measurement of the interior orientation is necessary then.

For the interior orientation at least four fiducial marks and the focal length are required. The nominal co-ordinates of the fiducial marks and the focal length, all in millimetres, must be provided—see the calibration certificate of the camera, or use standard values for Zeiss-RMK- or Wild-RC-cameras.

The button **Calibration data** opens another window, giving you possibilities to take the radial-symmetric lens distortion and the image principal point (PPS) into account:

- According to the formula $R * (K1 * R^2 + K2 * R^4 + K3 * R^6)$ (approach of BROWN)
- According to the formula $K1 + K2 * R + K3 * R^2 + K4 * R^3$ (LISA internal)

- Use of distortion values from BLUH (file SYSIM1.DAT)
- Data from a table.

Example for a table with distortion values:

```
0.0    0.000          each line: radius [mm], distortion [mm]
1.0    0.007
2.0    0.013
3.0    0.020
4.0    0.026          (etc.)
```

A click onto the **OK** button stores the selected parameters. If no distortion correction should be used, click onto the **Reset** button.

The specifications will be stored in a file having a CMR extension in c:\users \public\lisa\cmr. Example:

```
1   113.000      0.000          fiducial mark 1, x, y in [mm]
2     0.000  -113.000          fiducial mark 2, ...
3  -113.000      0.000          ...
4     0.000   113.000          ...
153.000                         focal length [mm]
DP   -0.9999990000E+00   0.0000000000E+00   distortion param.
DP    0.0000000000E+00   0.0000000000E+00
PP    0.0000000000E+00   0.0000000000E+00   principal point
CS   10.000   10.000    160                 pixel size, diagonal
```

If calibration data were used, additional the figure CALIB_1.IMA is created showing the graph of the distortion function.

Digital

Define the following parameters: Number of columns and rows of the sensor (landscape format, the bigger value = x!), pixel size in [μm] and the focal length in [mm]. If the pixel size is unknown, it can be calculated from the chip size in inch (e.g. 1/2.7″) or mm (e.g. 36 × 24). Or use a search engine and give in the name of your camera as well as the keyword "pixel pitch".

The programme creates two files, one defining the camera as previously described (file extension CMR), the other being universally valid displaying the parameters of the interior orientation. The latter has the same name as the camera definition file but the extension INN. The interior orientation process for each individual image described below is not necessary in this case.

Remarks to the images:

- For LISA, always and only use the original images created by the camera.
- If you want to rotate the images, do this only with LISA and not before with another software! This is important, for instance to make sure that calibration parameters are turned simultaneously.

10.7.4   Pre Programmes > Control Points

For creating or editing a control points file. Such a file is necessary for example if the exterior orientation should be defined by the measurement of at least four points per image. Aerial triangulation also requires a control points. For a maximum of 900 points.

Remark: In contrast to two-dimensional orientations and image rectifications like in LISA BASIC, in photogrammetry these options work three-dimensionally. For that, three-dimensional point co-ordinates (with z values) are necessary here!

10.7.5   Pre Programmes > Strip Definition

Many options like the automatic measurement of image co-ordinates for aerial triangulation (AATM) need information about the strips in the block. For each strip, the number of the first and the last image has to be defined; these numbers may have between 1 and 6 decimal digits.

The number of strips which can be defined here is limited to 20.

Example for the output file STRIP.DAT:

```
0          134 0        140    1     0.0000
0          155 0        161    1     0.0000
0          170 0        164    1     0.0000
```

(The format of this file is compatible to the programme BLOR).

10.7.6   Pre Programmes > Orientation > Measure >
Interior Orientation

This option is only necessary for scanned analogue images!

Important: To begin with take notice of the fiducial marks' position in relation to each other, respectively in relation to the side information bar. Example: If fiducial mark 1 is, according to the calibration certificate, placed in the middle of the left margin, this will relate to the original photo. Depending on the way the photo was placed on the scanner, fiducial mark 1 might appear rotated by 90° in the digital image, thus be positioned in the middle of the top end margin. This must be taken into account in the camera definition (see above)!

For every image to be processed, an interior orientation has to be carried out previously. After specifying the camera definition file, the fiducial marks defined there will automatically and successively be pre-positioned to their approximate values. The centre of each fiducial mark in question must be brought in line with the measuring mark using the middle mouse button depressed (or the Ctrl key or the arrow keys); to digitise the position finally click onto the left mouse button. Note: If the fiducial marks (usually little white dots) are hard to identify, it might be helpful to optimise the display using the brightness/contrast regulators. Points which cannot be measured can be skipped by clicking onto the right mouse button.

The option **Centre** activates an automatic centring. Therefore it suffices to hit the mark "more or less". This procedure may however only be applied in connection with point-shaped white marks! Alternatively, choose a rather great enlargement, then measure the fiducial marks manually as exact as possible, disregarding the option **Centre**.

For more than three fiducials a least squares adjustment is performed and the residuals in [mm] are displayed. This allows extreme values (peaks) to be marked and deleted from the calculation and fiducials to be measured anew. Finally click onto the **Ready** button—this will save the ascertained parameters.

For control, the calculated scan resolution in [dpi] as well as in [µm] will be displayed. If these values differ significantly from the real ones (chosen for scanning) the fiducial mark's nominal co-ordinates may be wrong.

The results will be saved in a file, carrying the same name as the image file but the extension INN. Example:

```
   0.1404250000E+04    0.1181858407E+02    transf. parameters
  -0.9734513274E-01    0.0000000000E+00    ...
   0.1399000000E+04    0.9734513274E-01    ...
   0.1175221239E+02    0.0000000000E+00    ...
 1      2740.000    1410.000               fiducial marks,
 2      1415.000      71.000               pixel co-ordinates
 3        69.000    1388.000               ...
 4      1393.000    2727.000               ...
RMK_1523.CMR                               camera defin. file
     153.000                               focal length [mm]
```

The transformation parameters refer to the transition from pixel to image co-ordinates.

10.7.7   Pre Programmes > Orientation > Measure > Exterior Orientation

If the results of a triangulation with BLUH are available, no exterior orientation needs to be executed—the parameters from the corresponding file (usually DAPOR. DAT) will be used. In case the parameters of the exterior orientation are known from a previous measurement they can be used (else click the **Reset** button).

For each control point the following steps are necessary:

- Select (mark) the point which shall be measured in the list below.
- Adjust the point by shifting the image with the central mouse button depressed or **Ctrl** or the arrow keys until the point and the measuring mark are precisely aligned one over the other.
- Digitising (by clicking with the left mouse button). The point and its number will be displayed in the image and marked with "M" in the listing below.

Note: It is a good idea to start with three or four well distributed, non-collinear points near to the image corners, in this way helping the orientation algorithm to converge.

To make it easier finding a point, optionally a neighbourhood of 121 × 121 pixels of the point can be stored. Choosing the option **Create point sketches**, the image part will be stored as a small image file. The file name has the form <point number>.QLK. If such a file already exists for a selected point, it will be displayed during the point measurement.

More than four control points produce an over-determination. A least squares adjustment and an indication of residuals with the option to mark and to delete points falling out of the defined limits will be carried out (button **(De)activate**). As a rule, delete as few points as possible and take care of an equal distribution of the points in the image. After four measured control points, any further point now will be pre-positioned in its approximate position.

Finally click the **Ready** button. The results will be saved in a file, carrying the same name as the image file but the extension ABS. Example:

```
     153.000                                         focal length [mm]
          .008           .006         1.587          φ, ω, κ [radians]
 1136701.547   970322.348      5289.731              X0, Y0, Z0 [m]
 120011 −108.016 70.005 2548514.900 5689958.100 38.200
 120072 −96.000   −8.455 2548720.500 5688872.700 41.600
 120122 −69.805 −66.654 2549108.300 5688075.100 31.200
 . . .
 . . .           (... image and terrain co-ordinates of all measured points)
 . . .
 CONTROL.DAT                                        (control points file)
      0.003       0.003                             (residuals in x und y [mm])
```

To control the results, please note the following:

- In case of aerial (vertical) images, the absolute values of φ and ω are usually less than 1.
- κ shows the flight direction—east having the value 0 and the angle is being issued counter-clockwise rotating, so representing north as ca. 1.57, west as ca. 3.14, south as ca. 4.71.
- The height of the projection centre (Z0) is the sum of the terrain- and the flight height.
- The standard deviation of the residuals at the control points should not be more than one pixel. The pixel size results from the photo scale and the scan resolution (see table in Sect. 3.2) or was defined within the camera definition (digital) as pixel size of the sensor.

10.7.8 Pre Programmes > Orientation > Measure > Calibration Pattern

Option for the calculation of lens distortion and principal point. Use for this our calibration pattern.

Display the calibration pattern on a sufficiently large flatscreen in full-image mode, for instance with the Windows image display. Now take a photo from the calibration pattern in the way that it nearly fills the whole image format (landscape format!). Make sure that all of the target marks are within the im-age. Also take care of a good illumination and a steady hand. Store the image with a numerical name (e.g. 1000.JPG).

Now start **Pre programmes > Orientation**, select the image from before and then go to **Measure > Calibration pattern**. The amount of target marks is 11 in x and 8 in y.

The programme starts near to the lower left image corner. Measure the first four target marks manually (lower left, lower right, upper right, upper left). Measure the first mark with special care, because from this position a small part of the image is stored and used as a reference for the following marks. The other marks will be measured automatically. At the end of the measurement, the calculation of the radial-symmetric lens distortion and the principal point starts. The result can be viewed in a graphics called CALIB_2.IMA.

The results of the measurement are stored in a file with the name of the camera and the extension CAL. Now start again the option **Pre programmes > Camera definition > Digital** and activate the option **Calibration data**. In the next window, just click onto **OK**.

10.7.9   Pre Programmes > Parameters of the Exterior Orientation

If the parameters of the exterior orientation are already known, they might be imported. The order of the angles φ, ω, κ during their calculation must be recognised—you may know, that the values of these angles depend on the sequence of their calculation! In LISA and BLUH the order is φ − ω − κ. If the angles were calculated in the order ω − φ − κ, please select the corresponding option.

Input data: Rotation angles in grads (full circle = 400 grads), degrees (full circle = 360°) or radians (full circle = 2π), co-ordinates of the projection centre in [m].

Option Roll-Pitch-Yaw: If data from the exterior orientation were collected during the flight (direct sensor orientation with GPS/IMU), the angles must be converted into the photogrammetric system φ, ω, κ. Input format: Image number, rotation angles (roll, pitch, yaw) in decimal degrees with North = 0 for yaw, projection centre (X0, Y0, Z0) in geographic co-ordinates (sequence longitude, latitude in decimal degrees, height in metres). Example:

```
4   8.981   1.498 173.291   9.874096 53.378387   235.880
5   7.252  -0.152 169.779   9.873602 53.378589   242.190
6   3.805   1.387 166.667   9.873115 53.378815   247.240
```

10.7.10   Pre Programmes > Select Model

If several models have already been defined (see next option), one of them may be selected here. Otherwise the latest active model will be used automatically. The model currently active is being indicated at the status line and stored in a file called STEREO__.PRD.

10.7.11   Pre Programmes > Define Model

Input parameters: Number of the left and the right image, method of exterior orientation:

- Parameters from BLUH, then the corresponding file must be provided containing the orientation parameters (default: DAPOR.DAT). In this case the file with the adjusted terrain co-ordinates (default: DAXYZ.DAT) may be used as object co-ordinates file.

- Parameters from ABS files, see Orientation > Measure > Exterior orientation for instance. In this case the control point's file may be used as **object co-ordinates file.**

As an option, all models of the block as given in the strip definition (see above) can be operated one by one (batch mode). If the input window already contains data of an existing model, the image numbers can simply be switched using the ◁ resp. ▷ buttons.

The stereo model comprises two kinds of parallaxes. The x-parallaxes are a result of the relief-induced radial-symmetric displacement and are necessary to determine the heights. Without calibration data there might be parallaxes of some pixels also in y direction, which afterwards might be disturbing during an automatic DTM generation (matching). The programme can minimize the remaining y parallaxes.

The programme calculates the co-ordinate range of the model in x and y. Important: Within the model area there must exist at least one point (from the object co-ordinates file)! If this is not fulfilled, go immediately to the stereo measurement and digitise some well-distributed points. For your information, some additional parameters are displayed:

- The approximate pixel size of the input images in terrain units (geometric resolution): This value can serve as point of reference for the pixel size in the project.
- The ratio distance/base: the higher this value, the less certain is the measurement z values.
- The maximum accuracy to be achieved in z depends on these parameters.

The data will be saved in a file whose name is constituted by the left and the right image number and which carries the extension MOD. Example:

135	136	Image numbers
DAPOR.DAT		File with orientations (*)
DAXYZ.DAT		Object co-ordinates file
1135300.000	1138000.000	Model range in x [m]
969300.000	971482.000	Model range in y [m]
0.0045		Relative angle
1		Parallax correction [pxl]

(*) If the orientations are taken from ABS files, this line keeps empty.

10.7.12   Aerial Triangulation Measurement (ATM)

Some pre remarks about the image and point numbers:

Image numbers: All images within the block must have a unique number! Concerning the names of the image files see File > Import.

Point numbers: Like before, also object points must have each a unique number! The automatic numbering in the manual or automatic measurement (see there) uses the image numbers and a consecutive index—for example, points within image No. 712 will get the numbers 712001, 712002, 712003 and so on. During the manual measurement of connection points (see there) numbers like 777770001, 777770002 etc. are created. This must be taken into account when numbering the control points! If, for instance, all images of the block have a three digit number, the control points may be named 1001, 1002, 1003 etc. without any conflicts with other object points.

10.7.13   ATM > Manual Measurement

With this module image co-ordinates can be measured for the aerial triangulation in BLUH. To do so a camera definition is necessary, furthermore the interior orientation of all images must exist.

Remark: Having good image material, an automatic measurement may be carried out instead (see below). But even then, the option described here has to be used to measure the control points or additional tie points. Per model, a maximum of 900 points can be measured.

Both image numbers, the approximate longitudinal overlap of the model ("endlap", mostly around 60 or 80 %) and the name of the output file must be provided. If the input window already contains data of an existing model, the image numbers can simply be switched using the ⊲ resp. ⊳ buttons. As already mentioned above within the exterior orientation, also here exists the option to store neighbourhoods of measured points as point sketches to help finding the exact position within further measurements.

If a file exists containing the orientation parameters and also a file with the ground co-ordinates of the control points, then these files may be defined. Then, measuring a control point, the corresponding positions in the left and the right image will be set automatically.

For technical reasons the models of one strip should always be worked on starting on the left proceeding to right. This means that for the first model the left and the neighbouring right image of a strip should be taken, then in the next model the former right becomes the current left image and so on.

Display of the images

The left and the right image of the stereo model can be displayed on the screen in two variations:

- Neighbouring left–right
- Overlaying each other, colour coded following the anaglyph method.

Trained users are able to see the first display mode in three dimensions. Less trained users should apply an alternative method, namely observe the situation through red-cyan glasses (red filter on the left side). The shape and colour of the measuring marks, under which the image parts are moved, may be altered using the corresponding buttons. An overview image with a rectangle showing the actual position facilitates the coarse positioning within the model. The image display can be performed in several sizes (zoom); the brightness can be regulated separately for the left and the right image.

Roaming in the model

The mouse executes the movement in x-y-direction; the central mouse button is to be held pressed down. Should any difficulties occur, the Ctrl button can be used instead. In addition, for precise positioning the arrow keys may be used, for fast positioning you can move the rectangular mark in the overview image.

The left and the right image are normally linked together. To shift the x- and y-parallax the right mouse button has to be held down. In this case only the right image will be moved. As soon as it is brought in line with the left image (parallaxes moved away), the programme may attempt to maintain the correct junction by permanent correlation while the images are moved: Choose the option Correlation.

Point measurement

There are four options to measure image co-ordinates:

- From previous model
- Gruber points
- Individual
- Strip connection

The registration of the image co-ordinates will be carried out with a click onto the left mouse button after the left and the right image part are set to corresponding positions. The options:

From previous model: Two cases are to be distinguished: (A) Points that have been measured previously in the present model, will be displayed coloured blue in the overview image and cannot be measured anew. Should a point be measured again it must be erased in this model before. (B) For points that have already been measured in the (now) left image of the actual model the programme will estimate considering the side lap, if they may possibly also be present in the right image. If this is the case then the programme will mark these in green in the overview image and will automatically set them in the left image; their position is fixed here (automatic point transfer). Accordingly, just the corresponding position in the right image is to be set manually. This option can and should be used from the second model onwards in the strip. If it is not possible to measure a point, the Skip button or the F3 key may be used.

Gruber points: To connect both images, at least 6 well-distributed points of the model have to be measured. From the second model onwards the three ones on the left side have already been measured in the previous model and can therefore be adopted. The default distribution is similar to the "six" on a dice, which means two points on the top, two in the middle and two on the bottom of the model. The programme sets those positions automatically and provides point numbers, which are extracted from an increasing index and the left image's number. Example: Left image number = 747, then the point numbers will be set to 747001, 747002, 747003, 747004, 747005 and 747006. In the case that a point cannot be measured, the button Skip may be used or alternatively the F3 key like before.

Individual: After entering its number a point will be checked for in the output file, to find out whether it has already been measured in the left image. If it has, a pre-positioning will be performed in the left image as described above (From previous model). Otherwise the point is to be adjusted freely in both image parts. If it turns out that the point cannot be measured after entering its number, the Skip button or the F3 key can be applied again.

Strip connection: If you have already measured connection points with the option ATM > Measure connections (see below) and created point sketches, these will be pre-positioned approximately and their point sketches will be shown bottom left.

Remarks concerning all measuring methods:

• The button Ready terminates the relevant module. Afterwards starting one of the Measure options continue the process.
• The Measure > End option (or the Esc key) causes the measurements to be stored and the module terminated.

Position and number of each point will be superimposed into both image parts; additionally they will be marked red in the overview image. Pixel co-ordinates are stored with the row co-ordinates being mirrored—the origin therefore lies in the left bottom corner. The first line of each model includes the image number, the focal length, the camera name and the rotation angle of the images. The next lines includes the fiducial marks (co-ordinates of the interior orientation), followed by the values point number, x left, y left, x right, y right for each point which was registered. The end of a model is indicated with −99.

For a further application in BLUH the file is to be exported after the completion of all measurements into the respective format with the help of the option ATM > Export BLUH.

Example for the output file:

```
135000136      153.000 RMK_1523.CMR                              1
         1  2735.016  1389.988  2739.972  1410.063
         2  1406.985    54.021  1414.941    71.033
         3    64.970  1376.988    68.940  1388.030
         4  1392.022  2713.022  1393.045  2726.955
   135001  1426.000  2551.000   585.000  2552.000
   135002  1426.000  1417.000   540.000  1417.000
   135003  1426.000   284.000   587.000   272.000
   135004  2500.000  2543.000  1765.000  2560.000
   135005  2598.000  1402.000  1856.000  1402.000
   135006  2620.000   284.000  1842.000   252.000
      -99
```

10.7.14 ATM > Calculate Strip Images

This option is a pre-requisite for the measurement of connection points like described in the next chapter and especially necessary if the block contains more than one single strip.

The strip definition (see above) must already exist. For each strip of the block a special image is calculated containing the single images in a size of each 800 by 800 pixels side by side in the sequence in which they form the strip. The name of the output image is derived from the number of the first and the last image. Example: First image No. 134, last image No. 140, then the output file has the name ST_134140.IMA. For a maximum of 20 strips and 50 images per strip.

10.7.15 ATM > Measure Connections

With this option, connections points (tie points) between neighbouring images and strips can be measured, serving as initial values for the manual or automatic measurement (AATM, see next chapter). If you want to use the connection points for the manual image co-ordinate measurement, the option Create point sketches must be activated!

Load one strip into the upper part of the window, the next strip which has a lateral overlap (side lap) with the first one into the lower part of the window. The strips now can be moved independently with the mouse, middle button depressed, and set to the start or ending position using the buttons $\boxed{\leftarrow}$ resp. $\boxed{\rightarrow}$. Now click onto Measure and define the name of the output file, default is TIEPOINT.DAT.

Digitise the first connection point by clicking the left mouse button in all images in which it appears, after that click onto the right mouse button (important to increase the point number!). Always begin in the upper strip! The point will be registered, marked in all images with a small red square and labelled with an increasing number. Now digitise, if necessary after moving the strips, the next point in all images, then press the right mouse button, and so on.

Measured points can be (de-)activated within the point number list. After a click onto the respective button, the point number is set to a negative value. A second click will reset the number to the initial value. Points with negative numbers (de-activated) will not be stored in the output file.

Finish the measurement with the **Ready** button.

Here some additional remarks:

- The control points necessary for aerial triangulation of course must be measured in the original images at highest resolution (**ATM > Manual measurement**, see above). However, the connection points measured as described here are only used as initial positions.
- If, as a result of a large number of well distributed control points, a sufficient connection between neighbouring strips is already given, the separate measurement of connection points may be not necessary. It is also possible to measure some connection points only in areas with few control points.
- If the block consists of only one single strip, but the images are of bad quality or very low contrast, you can often get better results from the automatic measurement if you measure some connection points manually.
- The more connection points are located in the block, the more stable the strip connection will be! As a basic rule, each model should have at least one common connection point with each neighbouring strip.
- And of course only those points can be used for connection, which appear in at least two neighbouring images (model) per strip.

Example of the output file:

```
777770001        210.000        250.000            134
777770001        115.000        255.000            135
777770001         26.000        254.000            136
777770001        211.000         56.000            155
777770001        121.000         49.000            156
777770001         19.000         59.000            157
777770002        227.000        242.000            135
777770002        140.000        242.000            136
777770002         41.000        246.000            137
777770002        240.000         37.000            156
777770002        139.000         48.000            157
777770002         42.000         53.000            158
    . . .
```

First column = internal point number, second column = x value, third column = y value (each times pixel co-ordinates, measured in the 800 by 800 pixel images), fourth column = image number.

10.7.16 ATM > Automatic Measurement (AATM)

For the processing of aerial images; if the images are located in different strips, these should be in a parallel arrangement.

The following steps must already be carried out: Camera definition, interior ori-entation of all images, manual measurement of the connection points as described before (optional) and the manual measurement of the control points. The threshold value for the correlation coefficient, the correlation window size and the number of iterations must be set. The control points files (image and object co-ordinates), the connection points file (optional) as well as the output file has to be defined.

In a more or less regular distribution connecting points will be searched automatically and also transferred into the following model, if possible. This is done using image pyramids to get even better results beginning with a coarse approximation.

In each image, points are searched within a regular grid of 30 by 30 squares. As a result, the maximum amount of points depends on the longitudinal overlap in %— for example, an overlap of 60 % will give a maximum of 60 % from 900 points = 540 points. At the beginning, within the squares areas of maximum contrast are searched. Then the homologous point in the right image is determined via correlation. When this work is done in the actual model, a plausibility control is carried out concerning the x and y parallaxes to delete obviously wrong points. After this, a second approach is made using the improved approximations, after that the programme goes on with the next model.

The output file (default name AATM.DAT) has the same format as described above (manual Measurement) and contains the pixel co-ordinates of all collected points. This file will be converted into the BLUH format afterwards. Now carry out the block adjustment with BLUH. Use the default names for the output files as suggested: DAPOR.DAT (orientations) and DAXYZ.DAT (object co-ordinates)!

Usually, the results can be significantly enhanced:

The strip connection was made up to now only with the manually measured connection points and possibly with the control points if these are located in more than one strip. But usually, much more points may be used for this purpose. Therefore start again the option ATM > AATM and select now the option Data from BLUH. Now the programme calculates for each object point the images in which this point shall occur, and establish a much better strip connection, leading to a more stable block. Finally, run BLUH again.

Remark: If, in seldom cases, additional connection points must be measured afterwards using the ATM > Manual measurement option, the file described

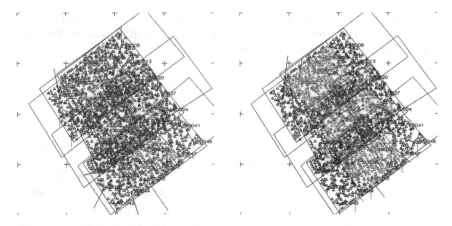

Fig. 10.1 Enhanced connection of the images (within each strip: *green dots*, between neighbouring strips: *red dots*)

before (standard AATM.DAT) should be used for output, choosing the option **Append** when the warning message "File already exists" appears. In this case, the export to BLUH must be started afterwards (Fig. 10.1).

The next step will be **Pre programmes > Model definition** with option **All** activated.

10.7.17   ATM > Export > BLUH

Input: File with the pixel co-ordinates from a manual or automatic measurement. The pixel co-ordinates are transformed onto the fiducial marks' nominal co-ordinates of the camera definition, becoming image co-ordinates. Example for the output file:

```
135000136     153.000
    13502       2.778      99.217     -68.507      98.679
    13503       2.238       2.836     -73.106       2.136
    13504       1.698     -93.460     -69.928     -95.321
    13505      93.678      98.093      31.333      98.531
    13506     101.429       1.076      38.225      -0.064
    13507     102.759     -93.954      36.238     -97.904
      -99
```

Note: This option is called automatically at the end of the AATM and so must not be started separately there.

10.7.18   Processing > Stereo Measurement

This module allows object co-ordinates to be measured in an orientated stereo model with an optional linked DTM. The camera definition, the orientation of both images as well as the model definition must already exist.

Note: If a DTM does not exist, it is possible to begin with an initial height which should be equivalent to the average object height. However, some of the following options become inapplicable then.

Display of the images

Identical with the display in the module ATM > Manual measurement (see there).

Roaming in the model

(a)   Object space → image space ("RRS", standard case)

The mouse executes the movement in x-y-direction; the central mouse button is to be held pressed down. Should any difficulties occur, the left and the right mouse button should be pressed down simultaneously OR the Ctrl button can be used.

Navigation in z direction (=elimination of the x parallax) is also mouse-controlled (wheel or right mouse button). While navigating over the model either the last fixed height may be maintained (option Z —), or the height of the connected DTM can be adopted permanently (option Z DTM).

The movement of the images is done by changing the x-y-z- co-ordinates in the object space. Using the collinearity equations, the corresponding image positions are calculated from these co-ordinates and the images are positioned to the measurement marks. So, the images are following the mouse movement.

(b)   Image space → object space ("RVS")

With the central mouse button pressed down both images are moved, with the right mouse button only the right image (similar as described above, see ATM > Manual measurement). From the actual image positions then the object co-ordinates are calculated with an intersection in space and displayed in the status line bottom left.

The Go to button allows a point to be positioned directly by manual input of its terrain co-ordinates. The Centre button resets the current position to the centre of the model. The button r = calculates the correlation coefficient in a 25 × 25 pixel neighbourhood for the current position of the left and the right image.

Modes of measuring

The registration of data is started by the Register option. Three-dimensional terrain co-ordinates can be captured (No., x, y, z). One of three alternative ways may be selected:

- Points/lines: Manual positioning and digitalisation of points and lines. As an option a subsequent pre-positioning is carried out regarding the points of an input file, provided they lie inside of the model range.
- Profile: The pre-positioning will be carried out after entering the starting and the final point as well as the step range (interval).
- Grid: The pre-positioning will be carried out after entering the co-ordinate area and the width of the raster; the defined area will be covered step by step beginning with the lowest x and y value.

Note: Pre-positioning has the effect that the x-y-values of the actual point cannot be changed—only the altitude can be set using the mouse, central wheel or right button depressed.

The digitalisation can be aborted with the Cancel button. If a pre-positioning takes place and a point cannot be measured stereoscopically there are two alternatives: Using the Skip button (or F3 key) causes the direct continuation of the process with the next point leaving the one in question unregistered. Alternatively, the z ? button (or the F4 key) stores the point with a z value of −999999.

Overlay > DTM points: This option is useful for example in combination with the stereo correlation (see next chapter). If the option Interpolation was deactivated and a more or less incomplete DTM created by matching, DTM points which are determined by correlation can be projected into the left and right image. In areas with big gaps, additional points should be measured manually (previous option Register) and subsequently this incomplete DTM and the file containing the additionally measured points should be combined into a complete DTM using the option Processing > DTM interpolation.

Overlay > Vector data: Projects the content of a vector file to be provided into both images (superimposition). Attention: For the transformation from terrain to pixel co-ordinates the z value from the vector file is used, not the height of an optional loaded DTM.

Overlay > New drawing: This undoes the options above by reading the images anew which might require some time.

And here one more idea: Eventually, you may be uncomfortable to work always with three changing mouse buttons—the central button for moving the model, the wheel or the right to set the elevation and the left to register the co-ordinates. For this, here are two suggestions:

- In any case, make sure to have a DTM, for example using the stereo correlation (see next chapter), and start the stereo measurement with it. Now it is not necessary to set the elevation.
- If you digitise points without a pre-positioning, just keep the Ctrl key depressed. With this, the model follows the mouse movement (like you would press the middle mouse button), and you only use the left mouse button for registration.

Single measurements

With the respective buttons you can measure distances, poly lines, angles and circle data (radius/centre).

10.7.19   Processing > Stereo Correlation (Matching)

The camera definition, the orientations of both images and the model definition must already exist. Only the actual model or all models of the block as given in the strip definition can be processed (batch mode), in the latter case the results will be joined to a mosaic afterwards.

In this module the elevations within the model area will be reconstructed. Because the orientation of each image is known, the programme can calculate corresponding pixel co-ordinates for any terrain point (x, y, z). As mentioned before, the process works "from bottom to top"—proceeding from an initial height z_0 in a given position (x, y), the z value will be modified until the resulting image parts fit ideally ("area based matching"). The criterion for this is the maximum of the correlation coefficient. In this way, a surface model (DTM) is generated.

The maximum displacement in x (px = x parallax) must be provided. Further, a threshold value for the correlation coefficient, the correlation window size and the number of iterations must be defined. A file with object co-ordinates of known points must exist. The following remarks concern the individual parameters:

px: Usual are values between 2 and 5. The stronger the relief, the higher must be this value. It should nevertheless not be set unnecessarily high to maintain precision and working speed because then wrong correlations may occur, for instance in areas with repetitive structures.

Correlation coefficient "r": In most cases the suggested value (0.7) can be maintained. Except for particular cases (e.g. low contrast images) it does not make much sense to choose a value of less than 0.6—this will cause more points to be correlated but also leads to a loss of accuracy.

Window: The greater the window, the more stable will be the results and the more time for calculation will be needed. Small correlation windows may lead to problems in areas with repetitive structures. You may start with a value between 7 and 13.

Filtering of the DTM: Please keep in mind that filtering will modify the height values especially at local minimum and maximum points. When a maximum of precision is of importance it is recommended to work without a filter. On the other hand filters have proven to be successful if for instance subsequently contour lines should be generated on the basic of the DTM.

Note: For the purpose of reassessment it may be advisable to deactivate the (standard) option Interpolation. The resulting DTM will then show gaps of different sizes, especially in areas with little contrast. In these areas points should be

measured manually—compare this also with the notes in the chapter **Processing >**
Stereo measurement as well as with the following chapter.

10.7.20   *Processing > DTM Interpolation*

In the case that a DTM generated by stereo correlation had shown gaps and
therefore additional points were measured manually, an area-covering DTM may be
interpolated with the help of this option from the initial DTM and the points
measured in addition. The option **model limits** restricts the interpolation to this area
instead of going to the project area borders of the DTM.

10.7.21   *Processing > Ortho Image*

This feature serves the purpose of the differential rectification of a digital image
onto an underlying DTM. The rectification quality depends on the quality of the
DTM! The images to be processed must be oriented entirely, and a DTM should
exist. As an alternative it is possible to rectify to a horizontal plane (z constant).

Three options concerning the input images are available:

- Single image
- Actual model
- All images

Single image: The image number must be defined as well as the source of the
exterior orientation: **From BLUH**, then also the file with the orientation parameters
(standard DAPOR.DAT), or via an ABS file for example from a manual
measurement.

Actual model: The model definition must have been carried out already. Then,
the left as well as the right image of the current model will being used in a way that
those features of the image lying closer to the left image's principal point will be
adopted from the left image; analogously on the right hand side (nearest nadir).

All images: Assume that a complete DTM already exists for the whole project
area (for example created within the stereo correlation, option **All**
models + Create mosaic used), then with this option all oriented images within
the working directory can be rectified and matched in one pass. The programme
takes all oriented images from the strip definition and process them one after the
other. The colour value definition within the ortho image will be done in the nearest
nadir mode.

In case of the second or the third option, a stepless colour value adjustment
between the particular images can be selected.

The rectification takes place on the area determined by the DTM or, in case of using a horizontal plane, within the project area. In positions for which no DTM information is available (blank spaces, e.g.), the ortho image remains empty. The geometric resolution (pixel size) of the ortho image is taken from the project definition.

10.7.22   *Processing > Camera Positions*

After the exterior orientations are defined for all images, a file with the camera positions can be created. These will either be taken from all ABS files located in the working directory or from the file DAPOR.DAT (BLUH).

10.7.23   *Processing > Image Sequences*

Pre-remark: This option is designed for automatic processing of image sequences (in this case stereo models photographed in a chronological order). The left image always corresponds with the same camera (position) and so does the right one. The image numbers must have 6 digits—then, the first digit refers to the camera number. Example: Image 100001 was taken from camera 1, image 200001 from camera 2. To make this simple, in the option File > Import you can connect selected or all images with chosen camera. By this, the reference to the camera definition files CAMERA_1.CMR and CAMERA_2.CMR can be handled.

About how to handle this tool: Beginning with the object points of the first model and a stereo correlation is executed. Object points are derived from the resolving DTM in a regular grid to serve as start values for the model definition of the ever next model to follow. Further, using the DTM and the images of the actual model, optionally an ortho image is created.

To make the programme track the images reliably a certain numbering mode (naming) of the image files is inevitable. Example with 10 models, then the image files may have the following names:

Model 1: left image 100001.IMA, right image 200001.IMA
Model 2: left image 100002.IMA, right image 200002.IMA
...
Model 10: left image 100010.IMA, right image 200010.IMA

In other words: All images have an unequivocal number, whereby the image numbers (-names) of the left and of the right camera are chosen in ascending order.

The exterior orientations of the first two images (first model) and the model definition of the first model must already exist.

Input data: Left and right image of the first model, left and right image of the last model. Finally, the input window with the parameters of the stereo correlation appears (see there).

Output:

- DTMs named GT_<left image, right image>, e.g. GT_10012001.IMA
- Optionally ortho image named OR_<left image, right image>, e.g. OR_10012001.IMA

10.7.24 Display Raster Image

Similar to the respective option in LISA BASIC.

10.7.25 Display Text

See the respective option in LISA BASIC.

10.8 Aerial Triangulation with BLUH

The programme system BLUH (bundle block adjustment) is developed and owned by Dr.-Ing. Karsten Jacobsen, University of Hannover, Germany. BLUH is also the necessary addition to LISA FOTO for the aerial triangulation of image blocks with the help of a bundle block adjustment.

For an easier handling in connection with LISA the programme BLUH_WIN was developed which controls the central BLUH modules from a Windows interface. For this, a parameter file called SYSTEM_BLUH.DAT is created and after that the selected module is started. The result lists of each module (extension LST, e.g. BLUH.LST) are displayed in the text editor after the module has finished.

The following descriptions refer mostly to BLUH_WIN. For more and detailed information about the programme system BLUH please use the descriptions which are delivered together with BLUH.

10.8.1 Pre Processing > Select Project, Define Project

See the respective options in LISA BASIC.

10.8.2   Pre Processing > Control Point Editor

See the respective option in LISA FOTO.

For the aerial triangulation with BLUH, for each control point a factor for the standard deviation in a range between 1 and 9 can be defined. Example: The standard deviation in BLUH was defined as 1 m for x, y and z. For a control point with problems in x and y, the factor may be set to 5, increasing the standard deviation for this point to 5 m. For x-y-control points the z-value and the factor for z must be set to 0, for z-control points the x- and y-value as well as the factor for x/y must be set to 0. Example:

```
80001 260834.230 9361733.530   868.000 1.00 1.00  all used
80002 261034.340 9367396.920   984.000 1.00 1.00
80003 261536.300 9369026.010   977.000 1.00 1.00
80004 261782.380 9369459.460   979.000 1.00 1.00
80005      0.000       0.000  1020.000 0.00 1.00  only z
80006 255501.000 9377104.000     0.000 1.00 0.00  only x, y
```

10.8.3   Pre Processing > Strip Definition

See the respective option in LISA FOTO.

10.8.4   Pre Processing > Photo Co-ordinates Editor

The data of the input file (usually DAPHO.DAT) are displayed in ascending point number order. After marking (clicking onto) a point in the list window, its number may be changed. The single point or all points with the same number can be erased using the respective button. Doing this, the number will initially be set to its negative value; the deletion can therefore be reversed by repeated marking and erasing (the point number will return to its original value).

Using the **OK** button causes the file to be actualised and stored; points with negative numbers will thereby be deleted.

10.8.5   Block Adjustment > Strategy

This is an option to help beginners setting the parameters. All you must know are some details like the approximate scale and the scan resolution (in case of analogue photos) or the approximate ground resolution and the pixel size of the sensor (in case of digital images) and some more.

Depending on the quality of the photo co-ordinates, the strip connection and the control points, several of the BLUH input parameters are set to useful values. Nevertheless, in the following modules you may change each of those parameters.

10.8.6   Block Adjustment > the Central BLUH Modules

Pre 1 (BLOR), Pre 2 (BLAPP) and Main (BLUH): See the respective BLUH manuals, also the description of BLIM (input parameters for the main programme).

BLOR

2D pre-check: Before starting BLOR, a 2D adjustment can be started to find large errors. Please remove the wrong data, then start Pre 1 (BLOR) again.

Remark: If BLOR will not run properly or with bad results it might be helpful to use the options Oblique photos > small view angle and Transformation > 2D. This will lead to a high stability even with not very good input data. It is also advised to use these options for images taken with a digital consumer camera, for instance operated on micro-drones (UAVs).

Automatic error correction: From the file BLOR.COR, created in the pre programme BLOR, optionally an error correction list named DACOR.DAT can be created and used in the following module BLAPP. Some remarks about this:

We have to divide between errors at control points, tie points of neighbouring strips and other points. The errors are listed in BLOR.COR according to their size in relation to the defined standard deviations, then marked with asterisks from no asterisks = small error to 4 asterisks = large error. For each point group can be selected from which amount of asterisks an error correction will be carried out (for instance, errors at control points beginning with ***). Please take into account that any automatic correction is a bit dangerous. If, for example, the limit for all groups of points will be set to the lowest value (*), then it can happen that a lot of points will be ignored in the adjustment what may have a negative influence to the strip connection and the stability of the block. Therefore it is suggested to start with the following limits: Control points ****, tie points ***, other points *. According to the fact that automatic aerial triangulation measurements (AATM, like those in LISA FOTO) usually will find a large amount of points, the value of the third point group (other points) can be set to a low value in most of these cases.

BLUH

A general remark if you have images coming from a non-calibrated digital camera: Within the main module BLUH you can activate the option Use of additional parameters > automatic reduction and then select at least parameter 9 (radial-symmetric lens distortion, see the BLIM manual for further details). In the camera definition of LISA FOTO, use the option Calibration data and then from BLUH, using the data from SYSIM1.DAT, see the programme description of LISA FOTO for details. After the distortion values are added to the camera definition file, image co-ordinates from LISA FOTO are corrected by them when you use the option ATM > Export BLUH. For further runs of BLUH it is then no more necessary to use additional parameters.

If you have a block of images with high stability (good overlap within each strip and between the strips, precise and well-distributed control points, precise image co-ordinates) then you can try to adjust the nominal focal length and to calculate the principal point. This can be done with the additional parameters 13 (focal length), 14 and 15 (principal point). In the same way like described for the radial-symmetric lens distortion, you should use then these parameters in LISA FOTO, camera definition (data from SYSIM1.DAT).

10.8.7 Block Adjustment > All (Batch)

Runs BLOR, BLAPP and BLUH directly one after the other. This option is useful if for instance after a previous run one or more control points were changed.

10.8.8 Block Adjustment > Analysis (BLAN)

See the BLAN manual for details.

A remark concerning the option Distance for neighbouring points: BLAN will include a listing of neighbouring (or identical) points in the output file BLAN.LST, depending on this value. Example:

```
NEIGHBOURED OR IDENTICAL POINTS
===============================

                          DX       DY       DZ   HOR DIST
     866211    874104    .031    -.073     .047      .079
     866301    874050   -.074     .065     .092      .098
     866355    874137    .084    -.045     .014      .095
     . . .     . . .
     MEAN SQUARE          .061     .062     .050        14
```

If you now start again Pre 2 (BLAPP), these points are added to the file DACOR. DAT (see description of the automatic error correction, above) in the following way:

```
874104      866211          0          0     1
874050      866301          0          0     1
874137      866355          0          0     1
. . .       . . .
```

If you now run again Main (BLUH), then each of these pairs of point is united which improves the strip connection and the geometric stability of the block.

10.8.9   Display Graphics

For the display of plot files from BLAN. These have the format HP-GL-1 and the extension PLT.

10.8.10   Display Text

See the respective option in LISA BASIC.

10.8.11   Some More Theory

In the following, some more information about the BLUH modules is given. This was taken from the BLUH manuals (Jacobsen 2007) and reduced to the options available for LISA:

BLOR: Approximate image orientations are computed by means of combined strips. The relative orientation, strip formation, transformation of strips together and the transformation to the control points are checked for blunders by *data snooping*. BLOR also creates an image number list in a sequence which leads to a minimised band width of the reduced normal equation system of the bundle block adjustment.

The image co-ordinates of the input file are stored in a direct access file with an index corresponding to the image number. The mean values of repeated measurements are used if they are within a tolerance limit. That means, there is no restriction to the sequence of the image co-ordinates in the input file and no limitation of the number of independent data sets of an image in the input file.

Corresponding to the image numbers in the strips, neighbouring images are transformed together by a similarity transformation. The shift in x and y and the rotation are used as start values for the relative orientation. The relative orientation is not identical to the relative orientation made during data acquisition for data check because the mean values of the image co-ordinates from all data sets are used.

After the relative orientation, model co-ordinates are computed. Neighbouring models are transformed by similarity equations based on tie points in the strip. The build-up strips are stored in a scratch file. The orientation of any strip is determined by the orientation of the first image in any strip. After finishing the strip creation, neighbouring strips are transformed together by a two-dimensional similarity transformation. Then the internal block system will be transformed again in two dimensions to the horizontal control points. This method for receiving approximate values for the image orientations will not lead to precise results but is sufficient as initial information for the block adjustment and is a very robust solution (see also Fig. 5.2).

BLAPP: This module is sorting the image co-ordinates in a sequence which is optimal for the bundle block adjustment. The measured image co-ordinates may exist in mono or stereo arrangement (LISA: Stereo). For the bundle block adjustment all observations for an object point have to be present at the same time. That means, the measurements have to be re-arranged by the object points. In addition the order of points shall cause a minimal bandwidth of the reduced normal equation system of the bundle block adjustment.

BLIM: This module handles all input parameters and temporarily files for the main module BLUH which is started immediately thereafter.

BLUH: This main module is a bundle block adjustment based on the collinearity equations. Observations are image co-ordinates and the control point co-ordinates. The interior orientation must be known at least approximately. Unknowns are the image orientations and the object co-ordinates. In the adjustment the square sum of the image co-ordinate corrections multiplied with the weight will be minimised. A blunder detection by robust estimators is possible. Since the collinearity equations are not linear, Newton's method is used for iterative computation, that means, approximate values for the unknowns are required as input. The approximate image orientations usually are determined by BLOR (see above). Based on the approximate image orientations BLUH is computing in the zero iteration approximate object co-ordinates based on the image orientations and the image co-ordinates.

The blunder detection in the programme system BLUH will be done in two steps. The first search with the method of data snooping will take place in the module BLOR. The detected blunders should be corrected or eliminated before starting the bundle block adjustment—therefore it is suggested to use the option Error correction (Sect. 10.8.6). The first run of a data set with BLUH should be done with blunder detection by *robust estimators*.

The robust estimators will reduce the weight of the defective observations, so finally the influence of blunders to the block adjustment is limited. The main problem in the data acquisition for block adjustment is the correct identification of tie points between neighbouring strips; within the strips the tie is usually correct. The robust estimators will reduce the weights of the observations of a strip tie point in such a case for one strip. That means, also the tie within one strip will be lost. If no re-measurement will be done, at least the point number of the observations within one strip should be changed—this will solve the discrepancy of the tie between the strips and within the strip there is still a connection.

For the blunder *identification* one more observation than for the blunder *detection* is required. In the case of a blunder in the base direction, the blunder can be detected with 3 images but three intersections are available—any can be correct. Therefore one more observation is required for the correct identification of the blunder. In the case of a blunder across the base direction with 3 observations, two will have an intersection, so the blunder can be identified.

Even with robust estimators not all blunders can be detected in one programme run if large and smaller blunders are mixed because large blunders can cause deformations of the block which will not be eliminated totally during iteration with robust estimators. If the control points are used with large standard deviations during adjustment with robust estimators, the block geometry can get too weak, so the number of iterations should be limited to 2. With error free control points a larger number of iterations with robust estimators is possible. If numerical problems are occurring, the block adjustment should be repeated with a smaller number of iterations with robust estimators.

To enable a simple analysis of errors in control points, the control points are computed at first as tie points, after this as horizontal control points (marked by CP in the output list) and finally as vertical control points (marked by CZ). If no image co-ordinate corrections are listed for the tie point but for the horizontal or vertical control point, the geometric problems are caused by the ground co-ordinates or the identification in any image. If image co-ordinate corrections are listed also with approximately the same size for the control point used only as tie point, the problems are caused by some image co-ordinate measurements.

In the case of error free control points, the ground co-ordinates of the control points are used without change. But there can be also differences in the image co-ordinates of the control points. For this reason, an intersection based on the adjusted image orientations will not lead to the input values of the control points. So also in this case there are differences between the original control point co-ordinates and the computed ground co-ordinates of the control points.

10.9   LISA FFSAT

10.9.1   Introduction

The programme LISA FFSAT is a digital photogrammetric workstation for stereo satellite data. The general handling is the same as in LISA FOTO. The differences are results of the different geometry: With FOTO, imagery from central-perspective cameras can be processed. Most cameras operated on satellites are line-scanners, meaning that we have central perspective only within a single line (across the flight direction) but a parallel projection from line to line (in flight direction).

Satellites which collect stereo imagery usually work in the way that one line is "looking forward", another "looking backward". Therefore, the "left image" in the traditional photogrammetric sense is created by the looking forward scan line, the "right image" some minutes later by the looking backward scan line.

From this it is obvious that a classic photogrammetric orientation of such images is impossible. But there is an alternative: Together with each image you can get a list of parameters, the so-called rational polynomial coefficients (RPC). These used in a special algorithm build by polynomials replace the collinearity equations of the traditional photogrammetry, and give an approximate exterior orientation which then must be improved using ground control points:

$$x' = \frac{P_1(x, y, z)}{P_2(x, y, z)} \qquad y' = \frac{P_3(x, y, z)}{P_4(x, y, z)}$$

with $P_n(x, y, z) = a_1 + a_2 * y + a_3 * x + a_4 * z + a_5 * y * x + a_6*y*z + a_7*x*z + a_8*y^2 + a_9*x^2 + a_{10}*z^2 + a_{11}*y*x*z + a_{12}*y^3 + a_{13}*y*x^2 + a_{14}*y*z^2 + a_{15}*y^2*x + a_{16}*x^3 + a_{17}*x*z^2 + a_{18}*y^2*z + a_{19}*x^2*z + a_{20}*z^3$

Image co-ordinates calculated from ground co-ordinates using RPCs (a_i). In total, 80 coefficients are used—each 20 for the polynomials $P_1 \ldots P_4$).

10.9.2 Image Sources

Usually different kinds (levels) of images are offered meaning the way of geometric and radiometric processing and enhancement. Make sure that you have *original* images, sometimes called "RAW", "level 1" or similar—in other words, images without any geometric rectification. Also make sure that you not only have the images but also the RPC data and some ground control point data for the orientation improvement.

10.9.3 File > Select, Define or Edit Project

See respective options in LISA BASIC.

10.9.4 File > Import

FFSAT can only handle 8-bit- resp. 24-bit-images in the IMA format. Import formats are BMP, JPG, TIF or RAW. For images with 16 bits resolution a conversion to 8 bits is done. In addition, a rotation by 90° clockwise is possible.

10.9.5 Pre Programmes > Sensor Definition

This option replaces the "Camera definition" of LISA FOTO.

The RPC's mentioned above will give approximate image co-ordinates from points with given geographic ground co-ordinates (x, y, z; x = longitude, y = latitude). Usually all data we process and want to create will be in a Cartesian system like Gauss-Krueger or UTM. They must be transformed internally into longitude/latitude, so please select the system, the ellipsoid, the zone and the zone width. If the images were taken from an area on the southern hemisphere please also select this option.

The results are stored in a file with the name of the sensor and the extension CMR. Example:

```
 1    2
20   34   6    0
 1
```

First line: Sensor, co-ordinate system (1 = GK, 2 = UTM, 3 = Geographic). Second line: Ellipsoid (20 = WGS 84), zone number, zone width in degrees, southern hemisphere. Third line: Images turned by 90°.

10.9.6 Pre Programmes > Orientation Measurement

Similar to the respective option in LISA FOTO.

As written before, the RPC data gives only an approximate exterior orientation of the images. It is necessary to improve this by the measurement of control points. The advantage is that the RPC approach will lead to a helpful pre-positioning, so you easily can find each control point in the image.

Like in LISA FOTO you can use existing orientation data (or more precise, bias-and-drift improvement data for the orientation given by the RPCs), furthermore you can create point sketches for each control point. For a good improvement at least 4 well-distributed control points are necessary. If you have less than 4 points or if you have selected this option, only a simple shift will be calculated.

The results are stored in a file with the name of the image but the extension ABS. Example:

```
 0.6306896938E+02   0.1000234108E+01   0.4484747673E-03
-0.3580213963E+02  -0.1995917036E-03   0.1001294288E+01
BANDA_RPC.TXT
```

First line: Coefficients for the bias-and-drift improvement of x. Second line: The same for y. Third line: Name of the file with the RPC data.

Remark 1: The method of improvement (bias-and-drift = affine, or a simple shift) depend on the number and distribution of the available control points, the sensor and other parameters. For instance, with Ikonos data normally a shift will be enough to get good results.

Remark 2: Usually the RPC data files have the extension TXT (Cartosat, Ikonos, GeoEye), RPB (QuickBird) or PVL (OrbView-3).

10.9.7 Processing > Stereo Measurement, Correlation, DTM Interpolation

See the respective options in LISA FOTO.

10.9.8 Processing > Ortho Image

See the respective option in LISA FOTO. Unlike there, you can select only one single image for the ortho rectification.

Appendix

LISA File Formats

Raster images (.IMA)

The header contains some common data (150 byte), flags (20 byte) and pointers (200 byte) and therefore always has a length of 370 bytes.

(a) Common image data (150 bytes)

Byte	Format	Meaning
1 ... 6	I6	No. of image rows
7 ... 12	I6	No. of image columns
13 ... 14	I2	Bits per pixel (radiometric resolution)
15 ... 29	F15.4	minimum x value (left image border)
30 ... 44	F15.4	maximum x value (right image border)
45 ... 59	F15.4	minimum y value (lower image border)
60 ... 74	F15.4	maximum y value (upper image border)
75 ... 89	F15.4	pixel size (geometric resolution)
90 ... 104	F15.4	minimum z value (DTM: min. terrain height)
105 ... 119	F15.4	maximum z value (DTM: max. terrain height)
120 ... 134	F15.4	factor of height scaling
135 ... 144	F10.3	focal length [mm], either 0.
145 ... 150		reserve

© Springer-Verlag Berlin Heidelberg 2016
W. Linder, *Digital Photogrammetry*, DOI 10.1007/978-3-662-50463-5

(b) 20 flags (20 bytes)

Flag No.	Byte	Format	Meaning
1	151	I1	0 = image, 1 = DTM
2	152	I1	
3	153	I1	rotation angle (n*90)
...	...	...	
20	170	I1	reserve

(c) 20 pointers (200 bytes, give the first byte of the corresponding data set)

Pointer No.	Byte	Format	Data set
1	171 ... 180	I10	begin of palette (8 bits)
2	181 ... 190	I10	begin of image data
3	191 ... 200	I10	
...	...	...	
20	361 ... 370	I10	reserve

(d) Palette (only in 8-bit images)

Contain the palette entries. For all grey values (0 ... 255) three entries for red, green and blue exist (intensities, 0 ... 255, format 3I1). Therefore, the palette all times has a size of 768 bytes and is located in byte 371 ... 1138.

(e) Image data

The image data begins in byte 371, if no palette is present, or in byte 1139 else (see also pointer 1 and 2). For true colour images (24 bit) the colour information is stored pixel by pixel (BIP) using the sequence BGR like in the Windows-internal bytemap format.

Vector data (.DAT)

Vector data in LISA consists of three-dimensional point co-ordinates which are stored line-wise in the sequence number, x-value, y-value, z-value.

The notations x, y, z refer to a Cartesian co-ordinate system with axes arranged mathematically like at Gauss-Krueger or UTM co-ordinates. Non-cartesian co-ordinates like the so-called Geographical co-ordinates (longitude, latitude) cannot be processed correctly and must therefore be transformed in advance (e.g. with Vector data > Projections*)!*

Besides the co-ordinates, code commands and end-of-line information will be stored in the vector file. Meaning of the codes:

1 – 3000	single point, general
3001 – 3100	like before, raster symbol
3501 – 3600	like before, vector symbol
4001 – 5000	for DTM, with:
4006	significant points, filter resistant
4007	delete start points for free-cut areas (*)
5001 – 8000	points on lines, general
9001 – 9999	for DTM, with:
9008	border polygon for free-cut areas (*)
9009	"smooth" break line, may be filtered
9010	"hard" break line, filter resistant

(*) the z value is without meaning.

Example:

```
5001   -999999.000   -999999.000      1.000
1000      1000.000      1040.000     20.000
1001      1267.800       807.450     17.000
1002      1600.311      1197.020     21.500
 -99       -99.000       -99.000    -99.000
```

Notes regarding the individual entries:

Point numbers   Positive integer numbers, maximum 10 digits.
X-, Y-, Z- values   Real numbers, maximum 12 digits, 3 decimal-digits.
Codes   Positive integer numbers, maximum 6 digits.

Note regarding the parameter code: Codes between 1 and 5000 represent single points, code between 5001 and 9999 lines. It is possible to access particular data within a certain file using codes if, for example, lines have been assigned individual codes following criteria like "field border", "street" etc.

Attribute data (.DBF)

Information which refers only to a few locations (e.g. owner of each plot or horizontal expansion of single soil samples) may be managed as attribute data. This is done in DBF files (format dbase IV) which can either be processed in data base programmes. The first field must contain the x-value and the second one the y-value of a referential point, for instance centre co-ordinates of each plot or location co-ordinates of soil samples.

Note concerning the termini: The field corresponds to the column, the data set to a line of a table.

Each field has a well-defined data type. There are different data types:

- Numeric (for numbers) with the sub-types integer or real
- Logical (true or false)
- Text (alphanumerical)

Attribute data can also be used within the raster image display (see **Display raster image > Overlay > Photos/Texts**): In this way you can overlay images or texts onto a geocoded image. The names of the files have to be defined in the DBF file (first field x-value, second field y-value, third field file name). The x-y co-ordinates determine the reference point in the terrain. Example:

```
1000.000        1040.000        test1.bmp
1267.800         807.450        test2.jpg
1600.311        1197.020        test3.tif
1104.200         973.100        test4.txt
```

Remark: For the use of other file formats (e.g. PDF, AVI) a suitable program must exist on the computer (media player, PDF reader).

Palettes (.PAL)

Four entries each row: Colour value, intensities red, green and blue, all entries between 0 and 255. Example:

```
  0    0    0    0
  1    2   51    1
  2    5   54    1
  3    7   56    2
  4   10   58    2
 . . .
255  255  255  255
```

Text reference (.TXT)

Special text file which can be used for example to create a legend. The pixel values are connected with textual information. Example:

The image LAND.IMA contains information on the land use in four classes. Then the file LAND.TXT can be created with the following structure:

```
 2      meadow
12      field
 7      forest
 8      settlement
```

Some LISA options create a text reference automatically.

GCP positions for tutorial 2

On the following pages you find all of the models used in tutorial 2 (Chap. 5), giving you the possibility to find the approximate positions of the GCPs.

The models are arranged in the order in which they forms the block, from left to right within each strip.

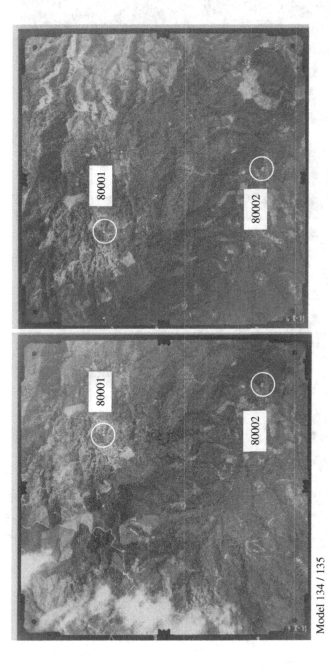

Model 134 / 135

Model 135 / 136

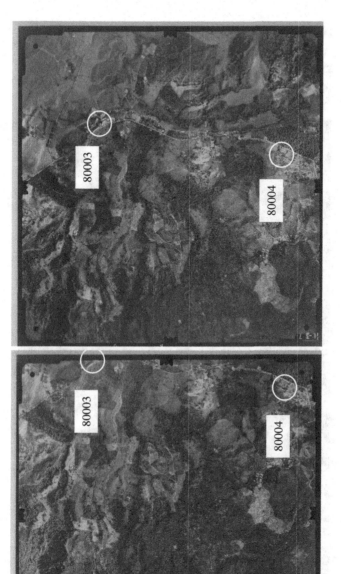

Model 136 / 137

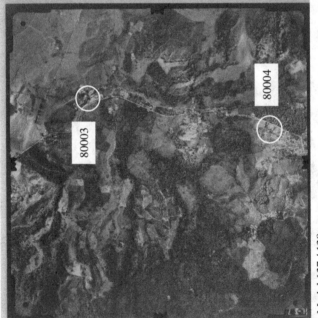

Model 137 / 138

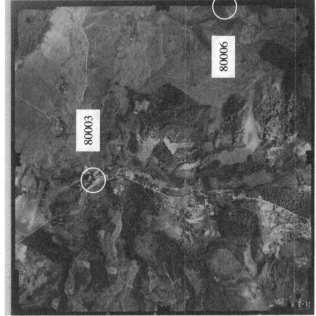

Model 138 / 139

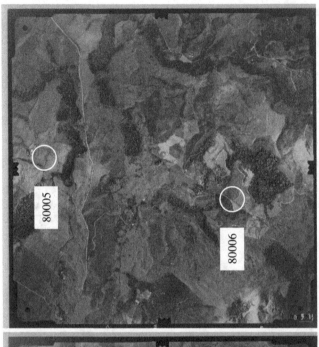

Model 139 / 140

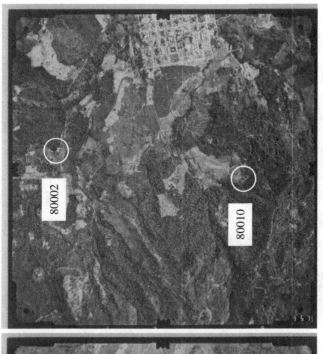

Model 155 / 156

Model 156 / 157

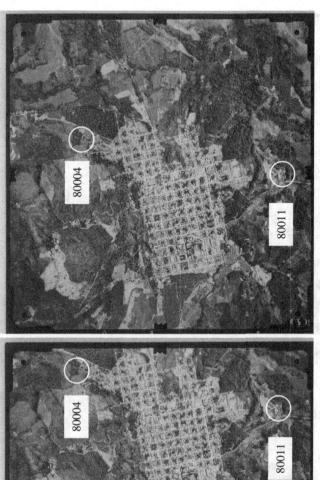

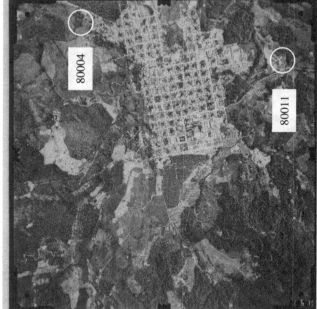

Model 157 / 158

Model 158 / 159

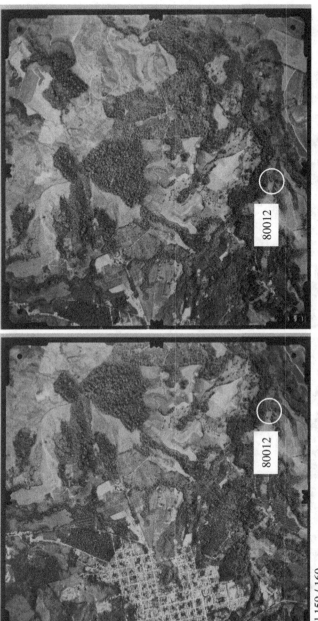

Model 159 / 160

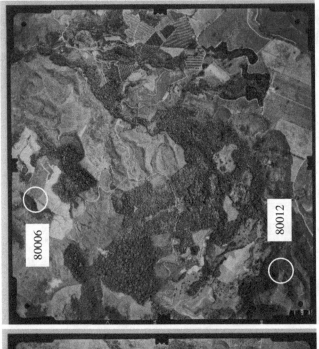

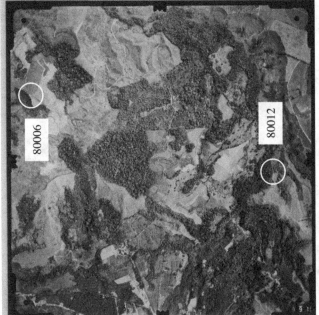

Model 160 / 161

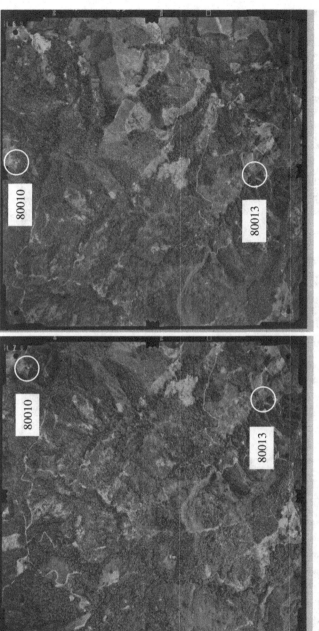

Model 170 / 169

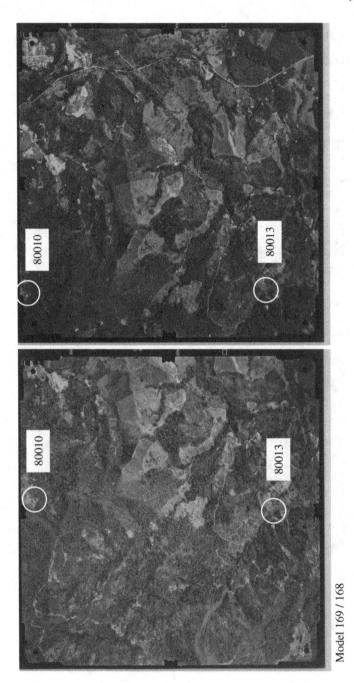

Model 169 / 168

Model 168 / 167

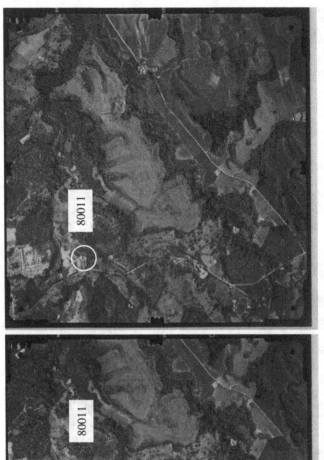

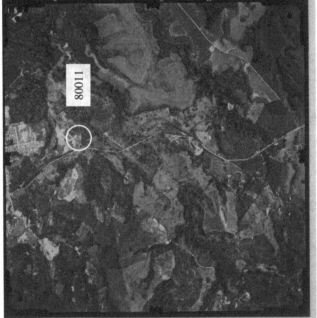

Model 167 / 166

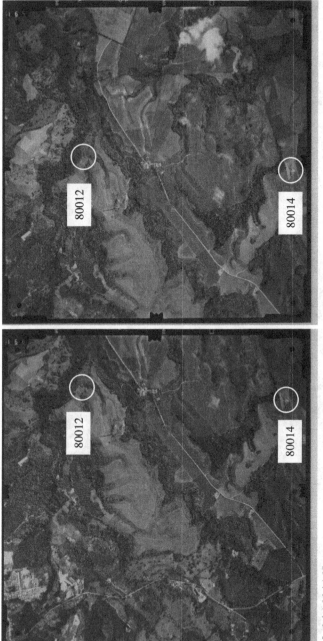

Model 166 / 165

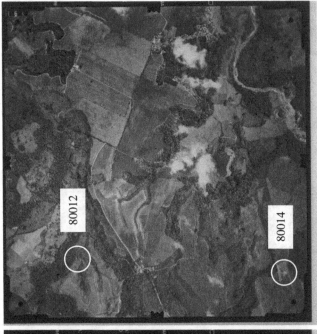

Model 165 / 164

Glossary

Anaglyphs Method of optical separation of the left and the right image for stereo viewing. Uses base colours as filters (red–green, red–blue or red–cyan). The images are displayed/printed overlaid, each in one of the base colour, and can then be viewed with a special spectacle

Base In Photogrammetry the distance between the projections centres of the left and the right image. See also → Height-base ratio

Block All images covering an area and being processed in a block adjustment, usually located in → strips

Calibration Method to calculate geometric or radiometric errors (distortions) of cameras or scanners. The results of a calibration can then be used to correct these errors

CCD Charge Coupled Device. Light-sensitive elements arranged in a line or in an area, used in digital cameras and scanners

Control points Points with known → object co-ordinates which can be found in an image, then used for instance to calculate the exterior → orientation

Correlation co-efficient Measure of similarity, used to compare two samples of data. The absolute value ranges between 0 (totally different) and 1 (identical)

DPI Dots Per Inch. Unit of geometric resolution of scanners, printers and other equipment. 1 in. = 2.54 cm

DSM Digital Surface Model. Describes the real heights of all objects (terrain, houses, trees, …)

DTM Digital Terrain Model: Describes only the terrain heights (without artificial objects)

Epipolar plane Defined by the projection centres of the left and the right image and the actual position on the object. Changing the height of the object will lead to a movement of the corresponding points in the images along epipolar lines

© Springer-Verlag Berlin Heidelberg 2016
W. Linder, *Digital Photogrammetry*, DOI 10.1007/978-3-662-50463-5

Fiducial marks In analogue → metric cameras used for the reconstruction of the interior → orientation of the photos. The marks define the → image co-ordinate system

Focal length Distance between the projection centre and the film plane (or the → CCD chip) of a camera, defines the opening angle

Frame camera Equipped with film or a → CCD area sensor. These cameras have a central perspective in contrary to systems with a line sensor

GCP Ground Control Point. See also → control points

Height-base ratio In aerial photogrammetry relation between the flying height above ground and the → base; equivalent in close-range photogrammetry is the ration distance/base. The value has a direct influence to the attainable accuracy of the calculated intersection of projection rays

Homologous points Object points which are located in two or more images from different positions. May be detected using the maximum of the → correlation coefficient

Image Here used for digital raster graphics, coming from a digital camera or a scanner

Image pyramids Set of images with decreasing resolution, used in image matching

Image space, image co-ordinates Two-dimensional co-ordinates measured in the images, units [mm]. In analogue cameras/photos the image co-ordinate system is defined by the → fiducial marks

Metric camera Camera with very high optical and mechanical precision, usually with fixed focal length, calibrated

Model In photogrammetry a pair of images taken from different positions. Also called stereo model

Nadir photo Aerial photo from a camera looking exactly down or in other words, the rotation angles φ and ω both have the value zero. Opposite: Oblique photos

Object space object co-ordinates, The terrain (aerial case) or in general the object (s) from which the images were taken. The co-ordinate system may be a "world system" like Gauss-Krueger or UTM but also can be a local one. Usually the co-ordinate axes are rectangular to another and the co-ordinates are given in metric units

Orientation The interior o. defines the relation between the camera and the image and can be calculated in the analogue case using the fiducial marks. The exterior o. defines the relation between the image and the → object space. Parameters of the exterior o. are the co-ordinates of the projection centre and the rotation angles

Parallaxes Co-ordinate differences of an object point in neighbouring images. The x parallax is a result from the camera positions and the relief, the y parallax is zero in an ideal case and should be corrected if not

Photo Here used for analogue images on film or paper in contrary to the digital representation ($\rightarrow$ image)

Pixel co-ordinates Pixel position of a point in rows and columns, counting from the upper left corner of a (digital) image

Radial-symmetric displacement Relief-depending displacements of objects in the image taken by a central perspective camera

Resampling Recalculation of grey values or colours in image processing. For instance necessary when an image shall be rectified

Residuals Remaining errors at $\rightarrow$ control points after an adjustment

Resolution The geometric r. is equal to the pixel size, the radiometric r. is equal to the number of grey tones or colours of an image

Stereoscopic viewing An image pair can be viewed stereoscopical (in a special manner) if the left image is only viewed by the left eye, the right image only by the right eye. One of several methods to achieve this are $\rightarrow$ anaglyphs

Strip In aerial photogrammetry all photos taken one after another in the same flight line

Surface model See $\rightarrow$ DSM

Terrain model See $\rightarrow$ DTM

Tie points Within an aerial triangulation used to connect models and strips

References

Albertz, J. & Wiggenhagen, M. (2005): Photogrammetrisches Taschenbuch / Photogrammetric Guide. 5th edition Heidelberg, 292p.

Bacher, U. (1998): Experimental Studies into Automatic DTM Generation on the DPW770. Int. Arch. of Photogrammetry and Remote Sensing, Vol. 32, Part 4, pp 35-41.

Baltsavias, P.B. & Waegli, B. (1996): Quality analysis and calibration of DTP scanners. IAPRS, Vol. 31, Part B1, pp 13-19.

Behan, A. & Moss, R. (2006): Close-Range Photogrammetric Measurement and 3D Modelling for Irish Medieval Architectural Studies. The 7th International Symposium on Virtual Reality, Archaeology and Cultural Heritage. Project presentations.

Brown, D. C. (1971) : Close range camera calibration. Photogrammetric Engineering, Vol. 8, pp 855-866.

Büyüksalih, G. & Li, Z. (2003) : Practical experiences with automatic aerial triangulation using different software packages. Photogrammetric Record, Vol. 18, pp 131-155.

Cáceres, B. et al. (2010): Evaluación geométrica del casquete glaciar del volcán Cotopaxi usando fotogrametría digital. Glaciares, nieves y hielos de América Latina. Cambio climático y amenazas. INGEOMINAS, Bogotá, Columbia, pp 225-238.

Cachão, M. et al. (2010) : Photogrammetric and spatial analysis of a bioeroded Early Miocene rocky shore, western Portugal. Online publication, Facies, Springer.

De Lange, N. (2002) : Geoinformatik in Theorie und Praxis. Heidelberg, Berlin, New York. 438 p.

Frick, W. (1995): Digitale Stereoauswertung mit der ImageStation. Zeitschrift für Photogrammetrie und Fernerkundung, H. 1, pp 23-29.

Gerwin, W. et al. (2009): Analyses of Initial Geomorphic Processes by Microdrone-based photogrammetry. American Geophysical Union, Fall Meeting 2009, abstract #EP31B-0591.

Grodecki, J. (2001): Ikonos Stereo Feature Extraction - RPC Approach. ASPRS annual convention St Louis.

Hannah, M.J. (1988): Digital stereo image matching techniques. International Archives of Photogrammetry and Remote Sensing, Vol. 27, Part B3, pp 280-293.

Hannah, M.J. (1989): A system for digital stereo image matching. Photogrammetric Engineering and Remote Sensing, Vol. 55, No. 12, pp 1765-1770.

Heck, V. & Vogel, S. (2009): Rectification of Historic Royal Air Force Aerial Photos and Generation of an Aerial Image Mosaic of the Sarno River Basin, Italy. Photogrammetrie, Fernerkundung, Geoinformation H. 3, pp 245-249.

Heipke C., (1990): Integration von digitaler Bildzuordnung, Punktbestimmung, Oberflächenrekonstruktion und Orthoprojektion innerhalb der digitalen Photogrammetrie, DGK-C 366, Beck'sche Verlagsbuchhandlung, München, 89 p (PhD thesis).

Heipke, C. (1995): State-of-the-art of Digital Photogrammetric Workstations for Topographic Applications. Photogrammetric Engineering and Remote Sensing, Vol. 61, pp 49-56.

Heipke C., (1995): Digitale photogrammetrische Arbeitsstationen, DGK-C 450, Beck'sche Verlagsbuchhandlung, München, 111 p.

© Springer-Verlag Berlin Heidelberg 2016
W. Linder, *Digital Photogrammetry*, DOI 10.1007/978-3-662-50463-5

Heipke, C. (1996): Overview of Image Matching Techniques. In Kölbl O. (Ed.), OEEPE -
Workshop on the Application of Digital Photogrammetric Workstations, OEEPE Official
Publications No. 33, pp 173-189.

Heipke, C. (2004): Some requirements for Geographic Information Systems: A photogrammetric
point of view. Photogrammetric Engineering and Remote Sensing, Vol. 70, No. 2, pp 185-195

Heipke, C. & Eder, K. (1998): Performance of tie-point extraction in automatic aerial triangulation.
OEEPE Official Publication No. 35, Vol. 12, pp 125-185.

Helava, U.V. (1988): Object-space least squares correlation. Photogrammetric Engineering &
Remote Sensing, Vol. 54, No. 6, pp 711-714.

Hobrough, G.L. (1978): Digital online correlation. Bildmessung und Luftbildwesen, Heft 3, pp
79-86.

Jacobsen, K. (2000): Erstellung digitaler Orthophotos. GTZ Workshop zur Errichtung eines
Kompetenznetzwerks für die Sicherung von Grundstücksrechten, Land- und
Geodatenmanagement. Hannover, 8 p.

Jacobsen, K. (2001): New Developments in Digital Elevation Modelling. GeoInformatics No. 4,
pp 18 - 21.

Jacobsen, K. (2001): PC-Based Digital Photogrammetry, UN/Cospar ESA-Workshop on Data
Analysis and Image Processing Techniques, Damascus, 2001, volume 13 of `Seminars of the
UN Programme of Space Applications', selected Papers from Activities Held in 2001, 11 p.

Jacobsen, K. (2006): Understanding Geo-Information from High-Resolution Optical Satellites.
GIS Development Asia Pacifica, pp 24-28.

Jacobsen, K. (2007): Programme manuals BLUH, RAPORIO and RPCDEM. Institute for
Photogrammetry and GeoInformations, University of Hannover.

Jacobsen, K. (2007): Comparison of Image Orientation by Ikonos, QuickBird and OrbView-3.
EARSeL. `New Developments and Challenges in Remote Sensing'. Rotterdam, pp 667-676.

Jordan / Eggert / Kneissl (1972): Handbuch der Vermessungskunde Bd. IIIa / 2. pp 104 - 108.

Jayachandran, M. (2003): DEM accuracy in analytical and digital photogrammetry. GIS
Development, Vol. 3, pp 33-38.

Kaufmann, V. & Ladstaedter, R. (2002): Spatio-temporal analysis of the dynamic behaviour of the
Hochebenkar rock glaciers (Oetztal Alps, Austria) by means of digital photogrammetric
methods. Grazer Schriften der Geographie und Raumforschung, Bd. 37, pp 119-140.

Keating, T. J. (2003): Photogrammetry goes digital. GIS Development, Vol. 3, pp 29-31.

Konecny, G. (1978): Digitale Prozessoren für Differentialentzerrung und Bildkorrelation.
Bildmessung und Luftbildwesen, H. 3, pp 99-109.

Konecny, G. (1984): Photogrammetrie. 4. Auflage, Berlin, New York, 392 p.

Konecny, G. (1994): New Trends in Technology, and their Application - Photogrammetry and
Remote Sensing - From Analogue to Digital. 13th United Nations Cartographic Conference,
Beijing, China, May 9-18, 1994 (World Cartography).

Konecny, G. (2002): Geoinformation. Taylor & Francis, London, 247 p.

Konecny, G. & Pape, D. (1980): Correlation techniques and devices. Vortrag zum XIV.
ISP-Kongreß Hamburg. IPI Universität Hannover, Heft 6, pp 11-28. Also in: Photogrammetric
Engineering and Remote Sensing, 1981, pp.323-333

Leberl, F. & Gruber, M. (2003): Aerial film photogrammetry coming to an end: Large format aerial
digital camera. GIM International, Vol. 17, No. 6, pp 12-15.

Linder, W. (1991): Klimatisch und eruptionsbedingte Eismassenverluste am Nevado del Ruiz,
Kolumbien, während der letzten 50 Jahre. Eine Untersuchung auf der Basis digitaler
Höhenmodelle. Wiss. Arb. d. Fachr. Vermessungswesen d. Univ. Hannover, No. 173, 125 p
und Kartenteil.

Linder, W. & Meuser, H.-F. (1993): Automatic and interactive tiepointing. In: SAR Geocoding:
Data and Systems. Karlsruhe. pp 207-212.

Linder, W. (1994): Interpolation und Auswertung digitaler Geländemodelle mit Methoden der
digitalen Bildverarbeitung. Wiss. Arb. d. Fachr. Vermessungswesen d. Univ. Hannover,
No. 198, 101 p.

Linder, W. (1999): Geo-Informationssysteme – ein Studien- und Arbeitsbuch. Heidelberg, Berlin, New York. 170 p.

Lohmann, P. (2002): Segmentation and Filtering of Laser Scanner Digital Surface Models, Proc. of ISPRS Commission II Symposium on Integrated Systems for Spatial Data Production, Custodian and Decision Support, IAPRS, Volume XXXIV, part 2, pp. 311-315.

Masry, S.E. (1974): Digital correlation principles. Photogrammetric Engineering Vol. 3, pp 303-308.

Mayr, W. (2002): New exploitation methods and their relevance for traditional and modern imaging sensors. Vortrag zur 22. Wissenschaftlich-technischen Jahrestagung der DGPF, Neubrandenburg.

Miller, S.B., Helava, U.V. & De Venecia, K. (1992): Softcopy photogrammetric workstations. Photogrammetric Engineering & Remote Sensing, Vol. 58, pp 77-84.

Miller, S.B. & Walker, A.S. (1995): Die Entwicklung der digitalen photogrammetrischen Systeme von Leica und Helava. Zeitschrift für Photogrammetrie und Fernerkundung, H. 1, pp 4-15.

Mustaffar, M. & Mitchell, H.L. (2001): Improving area-based matching by using surface gradients in the pixel co-ordinate transformation. ISPRS Journal of Photogrammetry & Remote Sensing, Vol. 56, pp 42-52.

Petrie, G. (2003): Airborne digital frame cameras – the technology is really improved! GeoInformatics, Vol. 6, No. 7, pp 18-27.

Petrie, G., Toutin, T., Rammali, H. & Lanchon, C. (2001) : Chromo-Stereoscopy : 3D Stereo with orthoimages and DEM data. GeoInformatics, No. 7, pp 8-11.

Plugers, P. (2000): Product Survey on Digital Photogrammetric Workstations. GIM International, Vol. 7, pp 76-81.

Redweik, P., Marques, F. & Matildes, R. (2008): A strategy for detection and measurement of the cliff retreat in the coast of Algarve (Portugal). In: C. Jürgens (Ed.), Remote Sensing – New Challenge of High Resolution, Bochum, pp 298 – 310.

Redweik, P. et al. (2010): Spatial Analysis of Trace Fossils for Paleogeographic Studies. Vortrag zur AGILE 2010, The 13th AGILE International Conference on Geographic Information Science.

Reulke, R. (2003): Design and application of high resolution imaging systems. GIS Vol. 3, pp 30-37.

Rieke-Zapp, D., Wegmann, H., Nearing, M. & Santel, F. (2001): Digital Photogrammetry for Measuring Soil Surface Roughness, In: Proceedings of the year 2001 annual conference of the American Society for Photogrammetry & Remote Sensing ASPRS, April 23-27 2001, St. Louis.

Rollmann, W. (1853): Zwei neue stereoskopische Methoden. Annalen der Physik, vol. 166, Issue 9, pp.186-187.

Ruzgiene, B. (2007): Comparison between Digital Photogrammetric Systems. Geodezija ir Kartografija, Vol. 33, Nr. 3, pp 75 – 79.

Ruzgiene, B. (2008): Some Aspects in Photogrammetry Education at the Department of Geodesy and Cadastre of the VGTU. Geodezija ir Kartografija, Vol. 34, Nr. 1, pp 29 – 33.

Ruzgiene, B. & Gecyte, S. (2008): Optimisation the workflow in a hybrid Digital Photogrammetric System for production of geodata from aerial photographs. 7th International Conference of Environmental Engineering, pp 1460 – 1466.

Ruzgiene, B. & Aksamitauskas, C. (2013): The use of UAV systems for mapping of built-up area. The International Archives of the Photogrammetry, Remote Sensing and Spatial Information Sciences, Vol. XL-1/W2, pp 349-353.

Santel, F. (2001): Digitale Nahbereichsphotogrammetrie zur Erstellung von Oberflächenmodellen für Bodenerosionsversuche. Diplomarbeit, Universität Hannover, 119 p.

Santel, F., Heipke, C., Könnecke, S. & Wegmann, H. (2002): Image sequence matching for the determination of three-dimensional wave surfaces. Proceedings of the ISPRS Commision V Symposium, Corfu. Vol. XXXIV, part 5, pp 596-600.

Sasse, V. (1994): Beiträge zur digitalen Entzerrung auf Grund von Oberflächenrekonstruktion. Wiss. Arb. d. Fachr. Vermessungswesen d. Univ. Hannover, Nr. 199, 227 p.

Schenk, T. (1999): Digital Photogrammetry, Volume I. Terra Science, Laurelville, 428 p.

Schenk, T. & Krupnik, A. (1996): Ein Verfahren zur hierarchischen Mehrfachbildzuordnung im Objektraum. Zeitschrift für Photogrammetrie und Fernerkundung, H. 1, pp 2-11.

Schneider, C., Schnirch, M., Casassa, C., Acuña, C. & Kilian, R. (2007): Glacier inventory of the Gran Campo Nevado Ice Cap in the Southern Andes and glacier changes observed during recent decades. Global and Planetary Change, Vol. 59, pp 87 - 100.

Seiffert, T. et al. (2009): Microdrone-based photogrammetry for water catchment monitoring. Vortrag zur Jahrestagung der Gesellschaft für Ökologie, Bayreuth.

Spoerl, H. (1933): Die Feuerzangenbowle. Düsseldorf.

Thomas, H. & Cantré, S. (2009): Applications of low-budget photogrammetry in the geotechnical laboratory. Photogrammetric Record, No. 24, pp 332-350.

Usery, E. L. (1993): Virtual stereo display techniques for three-dimensional geographic data. Photogrammetric Engineering and Remote Sensing, No. 12. pp 1737-1744.

Walker, A.S. & Petrie, G. (1996): Digital Photogrammetric Workstations 1992-96. ISPRS congress Vienna. International Archives of Photogrammetry and Remote Sensing, Vol. XXXI, part B2, pp 384 – 395.

Wegmann, H., Rieke-Zapp, D. & Santel, F. (2001): Digitale Nahbereichsphotogrammetrie zur Erstellung von Oberflächenmodellen für Bodenerosionsversuche. Publikationen der DGPF, Band 9, Berlin.

Wiggenhagen, M. (2001): Geometrische und radiometrische Eigenschaften des Scanners Vexcel UltraScan 5000. Photogrammetrie, Fernerkundung, Geoinformation H. 1, pp 33-37.

Willkomm, P. & Dörstel, C. (1995): Digitaler Stereoplotter PHODIS ST – Workstation Design und Automatisierung photogrammetrischer Arbeitsgänge. Zeitschrift für Photogrammetrie und Fernerkundung, H. 1, pp 16-23.

Wundram, D. & Löffler, J. (2007): Kite Aerial Photography in High Mountain Ecosystem Research. Grazer Schriften der Geographie und Raumforschung, Band 43, pp 15 – 22.

Wundram, D. & Löffler, J. (2008): High-resolution spatial analysis of mountain landscapes using a low-altitude remote sensing approach. International Journal of Remote Sensing. Vol. 29, H. 4, pp 961 – 974.

Wrobel, B. & Ehlers, M. (1980): Digitale Korrelation von Fernerkundungsbildern aus Wattgebieten. Bildmessung und Luftbildwesen Nr. 48, pp 67-79.

Zhang, B. & Miller, S. (1997): Adaptive Automatic Terrain Extraction. Proceedings SPIE Vol. 3072, pp 27-36.

Index

A

AATM, 72, 76, 81, 82
ABM, 46
Absolute orientation, 30, 80, 122
Accuracy, 4, 11, 21–23, 29, 32, 34, 39, 40,
  50–53, 58, 59, 67, 76, 81, 200
Aerial camera, 4, 18
Aerial triangulation, 12, 19, 31, 61, 62, 67, 76,
  166, 201
Anaglyph method, 42, 59, 123
Analogue, 6, 28
Analysis, 76, 84, 172
Analytical plotter, 6, 11, 23
Angle, 21
Area, 127
Area filling, 127
Area sensor, 4, 109, 200
ASCII, 34, 83, 125, 140

B

Base, 11, 12, 21, 25, 32, 34, 40, 42, 96, 123,
  124, 172, 199, 200
Batch mode, 88
Bilinear, 54, 55
Block, 12, 13, 19, 23, 61–64, 67, 68, 75, 76,
  78, 79, 81, 83, 84, 88, 122, 170–172, 181,
  199
Block adjustment, 75
Blunders, 76, 77, 80, 170–172
BMP, 23, 24
Border, 26, 32, 40, 42, 44, 53
Break lines, 43
Brightness, 29, 31, 47, 55, 71

C

Calibration, ix, 5, 23, 26, 27, 75, 76, 199
Calibration pattern, 151
Camera definition, 26, 28, 94, 101, 146, 159,
  161, 163

Cartesian, 93
CCD, 6, 21, 40, 94, 199, 200
Central perspective, 1, 4, 6, 8, 40, 52, 200, 201
Central projection, 53
Close-range, ix, xii, 3, 4, 76, 99, 120
Code, 43, 58, 99, 105
Collinearity equations, 36, 39, 42, 76, 171
Colour value adjustment, 55
Connection points, 61, 73, 87
Contour lines, 138, 139
Contours, 32, 56, 57, 89–91, 97
Contrast, 11, 29, 31, 47, 48, 50, 59, 73, 96, 98,
  104, 106
Control point, 32, 148
Coordinates, 125
Correlation, 50
Correlation coefficient, 46–50, 104, 200
Correlation window, 49, 50, 73, 105
Cubic convolution, 55

D

Data collection, 58
Data reduction, 57, 76
Data snooping, 77, 170, 171
Differential DTM, 93, 96–98
Digital camera, 5, 6, 12, 18, 40, 94, 104, 200
Digitising, 43, 150
Displacements, 8, 9, 47, 52, 109, 201
DSM, 46, 53, 56
DTM, 43–45, 49–53, 56–58, 67, 88–91, 93,
  95–98, 101, 104–106, 121, 153, 161–165,
  199, 201

E

Endlap, 12, 23, 65, 73
Epipolar, 2, 40, 42, 47, 199
Equidistance, 57, 91
Error, 82
Error correction, 77, 79, 80, 82, 83

© Springer-Verlag Berlin Heidelberg 2016
W. Linder, *Digital Photogrammetry*, DOI 10.1007/978-3-662-50463-5